AF522283

Modern Fishing Gear Technology

Modern Fishing Gear Technology

George Hadwin

RANDOM PUBLICATIONS
NEW DELHI (INDIA)

Modern Fishing Gear Technology

ISBN 978-93-5111-931-9

Published in 2016 in India by

RANDOM PUBLICATIONS

4376-A/4B, Gali Murari Lal, Ansari Road
New Delhi-110 002
Phone : +9111-43580356, 011-23289044, 011-43142548
e-mail: sales@randompublications.com,
info@randompublications.com, randomexports@gmail.com

Reprinted 2025

Type Setting by : Friends Media, Delhi-110089
Digitally Printed at: Replika Press Pvt. Ltd.

Preface

Modern methods of catching fish all have analogues in traditional practices that have been used for thousands of years. As in modern fishing, the methods used historically were chosen because they were well suited to catching the targeted species. Spears and arrows, for example, were used when the target was a large fish that could be closely approached, either by stealth or by confining the fish in a trap of some sort.

A diverse range of fishing gears and methods have been evolved to capture a wide range of species constituting the fisheries in different parts of the world. Significant developments have taken place in the fishing gear materials, design and fabrication and degree of sophistication in the operation and handling of fishing gears, in the last three decades.

The introduction of any new fishing technology always demands good national management and regulation. Vessels must also be matched with new methods or gear. As gear becomes more complex, it may require upgrading of vessels in size, power, and design. The site specificity of fishing arts should always be considered.

Fish capture technology encompasses the process of catching any aquatic animal, using any kind of fishing methods, often operated from a vessel. Use of fishing methods varies, depending on the types of fisheries, and can range from a simple and small hook attached to a line to large and sophisticated midwater trawls or purse seines operated by large fishing vessels. The targets of capture fisheries can include aquatic organisms from small invertebrates to large tunas and whales, which might be found anywhere from the ocean surface to 2 000 meters deep.

This publication brings together latest information in the field of commercial fishing technology and provides an overview of the important fishing gears, their design, fabrication and operation with focus on recent advances in responsible fishing.

– Author

Contents

1

Introduction

DEFINITIONS

Methods to catch fish and other aquatic resources, with or without a gear, have always been practiced. Although the fundamental principles, *i.e.* filtering the water, luring and outwitting the prey and hunting, are the basis for most of the fishing gears and methods used even today, gears and methods have changed significantly over time and their capture efficiency is obviously hardly comparable to that of prehistoric times.

A fishing gear is the tool with which aquatic resources are captured, whereas the fishing method is how the gear is used. Gear also includes harvesting organisms when no particular gear (tool) is involved. Furthermore, the same fishing gear can be used in different ways. A common way to classify fishing gears and methods is based on the principles of how the fish or other prey are captured and, to a lesser extent, on the gear construction.

1. Surrounding nets (including purse seines)
2. Seine nets (including beach seines and Boat, Scottish/Danish seines)
3. Trawl nets (including Bottom: Beam, Otter and Pair trawls, and Midwater trawls: Otter and Pair trawls)
4. Dredges
5. Lift nets
6. Falling gears (including cast nets)
7. Gillnets and entangling nets (including set and drifting gillnets; trammel nets)
8. Traps (including pots, stow or bag nets, fixed traps)
9. Hooks and lines (including handlines, pole and lines, set or drifting longlines, trolling lines)
10. Grappling and wounding gears (including harpoons, spears, arrows, etc.)
11. Stupefying devices

This classification is being slightly modified to accommodate the most recent development of fishing gears and methods and will soon be published.

Fishing methods have continuously evolved throughout recorded history. Fishers are inventive and not afraid of trying new ideas. The opportunities for innovation have been especially good in recent decades with advances in fibre technology, mechanization of gear handling, improved performances of vessels and motorization, computer processing for gear design, navigation aids, fish detection to mention only a few technologies.

Whereas technological development of fishing gear and methods in the past was aimed to increase production, the present situation with many overfished stock, limited possibilities to expand fishing on underexploited resources and concerns about the environmental impact of fishing operation, gear development is now very much focussed on selective fishing and gears with less impact on the environment.

Artificial reefs are structures placed on the sea bottom to gather fish. They are either used alone or with Fish Aggregating Devices (FADs). Most artificial reefs are large, permanent structures set in rather shallow water. Generally they are made of modern synthetics and hardware (concrete, metal, plastic pipes) which are sometimes finished using local vegetal materials (bamboo, coconut, leaves, coir for the anchor rope, etc.). In some cases lighter structures made from local materials are installed temporarily to lure fish to a specific area, during a certain fishing season.

Artificial reefs are established for various purposes:

- To enhance resources in coastal waters in order to facilitate exploitation;
- To create a biological reserve; and,
- To prevent the use of certain fishing gear, particularly bottom trawls, in a given area.

CONSTRUCTION AND MAINTENANCE

Different structures can be used as artificial reefs, such as wrecks, offshore oil rigs and pipelines on the sea bottom, heaps of oyster shells, etc. as well as specifically designed modules. Constructions covering a large area can be composed of individual modules, of which there are many models depending on the purpose. For instance, to prevent trawling, a model might consist only of slabs of concrete with stakes. To enhance resources, that is to provide shelters for certain fishes and fish aggregation in general, modules should include holes (big enough to avoid being quickly sealed by marine organisms) where the material can facilitate the fixing of "fouling" organisms as food for aggregating fish.

The overall dimensions of the reef and numbers and sizes of the holes for shelter are critical factors, as well as waves around it (modules should not bury themselves in bottom sediment). The modules should be bulky enough to project from the ocean bottom and massive enough to remain in place (even during severe storms); it might be suitable to anchor the modules of the artificial reef, particularly in shallow waters.

Material should resist rapid corrosion and should not introduce harmful substances into the marine environment.

Various elements can be used to create artificial reefs, including:

- Scrap material (car, old ships and barges, scrap concrete, household appliances, rock and rubble from excavations, etc.);
- Baled urban solid waste;
- Products such as sandbags and bamboo frames to construct temporary "reefs";
- Large concrete blocks, elements made from FRP and PVC; and,
- Offshore structures for oil or gas exploitation can also serve as artificial reefs and fish aggregating devices.

Building costs and installation factors (a very large number of artificial reefs must be set up) are also considerations in the choice of modules used. In shallower waters, setting FADs in conjunction with an artificial reef has proven profitable in several places. It has not been possible to identify a single model which could be universally recommended.

AGGREGATION PROCESS

An artificial reef plays the role of an offshore bank and protected area or park. The process of aggregation is often made in several steps: concretizing the blocks, then aggregating small fish and finally attracting larger fish.

The first few fish species arrive and eat organisms which have fixed on the artificial reefs which act as a food reservoir for demersal fish. Then, other species, such as predators of the former come.

For certain species (rock fish, octopus, crustaceans) artificial reefs are shelters; for small fish/juveniles, reefs provide protection from predators. Artificial reefs are also spawning areas for certain species (*e.g.* cephalopods) which explains how artificial reefs can enhance resources. In shallow waters shoals of small pelagic fish have been observed to use artificial reefs as stopping grounds.

A typical school of the sparid, *Montaxis grandoculis* (Mu), circling an artificial reef. This fish primarily eats molluscs.

HISTORICAL FISHING METHODS

Modern methods of catching fish all have analogues in traditional practices that have been used for thousands of years. As in modern fishing, the methods used historically were chosen because they were well suited to catching the targeted species. Spears and arrows, for example, were used when the target was a large fish that could be closely approached, either by stealth or by confining the fish in a trap of some sort. Poisons made from fruits or bark were used to kill many fish simultaneously if the fish were physically confined. Archaeological evidence makes it clear that many types of traps were used extensively in traditional fishing.

Active traps included baited box traps and hooks. Passive traps included weirs, which were typically fence-like structures made of a row of wooden stakes that had latticework woven in between them. They were positioned in estuaries, streams, and shallow rivers to intercept migratory fish (*e.g.*, salmon traps). Other passive traps included coastal fish ponds in which small fish were raised to a suitable size in a simple form of aquaculture (*e.g.*, Hawaiian fish ponds). Nets made from natural fibers date back to the Stone Age. Some were used in an active mode, *e.g.*, beach seines. Others were used passively, *e.g.*, gill nets. All of these traditional fishing methods are relatively unimportant in modern commercial fisheries. They are, however, still used in subsistence fishing and in near-shore and local fisheries.

MODERN ANALOGUES

The modern analogue of the spear and arrow is the harpoon. Actually the use of hand-thrown harpoons to catch whales dates back at least to the 12th century, when the Spanish were hunting Right Whales in the Bay of Biscay. Modern-day harpoons, fired from a cannon and armed with an explosive head, are used today by Japan, Greenland, Norway, and South Korea to kill roughly 1,000 Minke whales per year in so-called "scientific whaling".

The present International Whaling Commission moratorium on virtually all commercial whaling reflects the ability of canon-fired harpoons with explosive heads and a number of other technological developments in the whaling industry to decimate the once numerous stocks of large baleen and sperm whales.

Hooks and lines are still used to catch fish, but the infrastructure surrounding their use in commercial fishing is very different now than in the past. Hand lines are used in trolling, which involves dragging a baited hook through the water behind a boat. Fish caught by trolling include mackerel and some game fish.

Hand lines are also used in conjunction with live bait in a procedure called chumming. Live bait are thrown into the water to induce a feeding frenzy. The target fish strike at the hooks, which are sometimes baited and sometimes hidden in the feathers of artificial lures.

When the target fish are "hot" (*i.e.*, sufficiently excited by the presence of the chum) they will even strike at bare hooks. Examples of fish caught by chumming include skipjack and albacore tuna. The procedure is limited by the availability of live bait, which must be maintained in a viable condition in bait wells aboard the fishing vessel. Fresh seawater is constantly circulated through the bait wells. In the case of tuna, about one kilogram of chum is needed to catch 10 kilograms of tuna. Trawl lines are another form of hook-and-line fishing. Trawl lines are very long lines (*i.e.*, a kilometer or more) to which hundreds of shorter lines with baited hooks are attached. Laid along the bottom, trawl lines target demersal fish and were once used exclusively to catch groundfish off the coast of New England. Cod, haddock, hake, halibut, perch, groupers and snappers are all targets of trawl line fishing. The technique is particularly useful where the bottom is too rugged for the use of nets.

Pelagic long lines are the analogue of trawl lines in the water column. The lines are supported by buoys and are set to depths ranging from roughly 50 to 250 meters. Pelagic long lines are much longer than trawl lines, with the main line being perhaps 60 to as much as 180 kilometers long. The main line is actually composed of 400 to 450 sections, with each section having a length of 150 to 400 meters.

Typically five branch lines with hooks form one section. As many as 2,000 hooks can be operated in one set. Paying out the line from the stern of the boat requires about four hours and usually starts before sunrise. Hauling in the line begins in the early afternoon and is facilitated with the use of a line hauler. The retrieval process can take 10 hours or more. Catch rates are typically 3-4 fish per 100 hooks. The crew's work is not finished until the catch has been prepared and sorted, and a work day can easily last 18 hours. The long work day combined with the fact that pelagic long liners stay at sea more than 200 days per year has made finding crews difficult. Yellowfin tuna account for about half and albacore tuna about one-third of the fish caught with pelagic long lines. Other fish caught by this method include bigeye and bluefin tuna, sharks, marlin, swordfish, and sailfish. In the case of swordfish, which feed at night, the lines are set in the afternoon and picked up the next morning.

Traps account for very little of the present day world fish catch, but they are important in certain fisheries. Lobster pots and crab pots, for example, are routinely used to catch lobsters and crabs, respectively. They are basically an enclosed framework of wire that can be easily entered by the target animal but from which escape is prevented by a retarding device. The bait consists of meat scraps such as fish parts or chicken necks. Placed on the bottom, the pots must be pulled and checked on a regular basis since the target animal, once trapped, has nothing to eat. Alternatively, if more than one crab or lobster enters the pot, the animals may become cannibalistic.

A variation on more traditional traps is the fish aggregation device or FAD. FADs are basically floating objects and typically amount to little more than a simple float moored to the sea floor, in some cases in water several thousand meters deep. Many fish are attracted to floating objects, and it is not uncommon to find numerous fish congregating beneath or near them. Why fish are attracted to FADs is a topic of debate, and the answer may be that not all species are attracted to FADs for the same reason. Tuna, for example, tend to ignore FADs unless prey are present. Other fish, such as mahimahi, seem to be attracted to the FAD itself. By mooring FADs at well defined locations and perhaps equipping them with radio transmitters, governmental agencies can greatly facilitate the job of finding pelagic fish. Mahimahi, tuna, and billfish are among the fish whose catch has been facilitated with FADs.

Weirs and pound nets continue to be used to capture finfish along the shoreline. Pound nets, for example, are used in the Mediterranean to catch tuna, along both coasts of the North Pacific for salmon, and in the western Baltic Sea to capture eels and herring. The word pound reflects the idea that the nets impound the fish, which are easily removed from the confinement of the crib. Interesting variations on the more traditional traps are the aerial traps used to catch flying fish. Once the fish have been stirred up, they are caught in the air with special gear called veranda nets. This technique is used, for example, by South Pacific islanders, who catch flying fish at night by attracting them with torches.

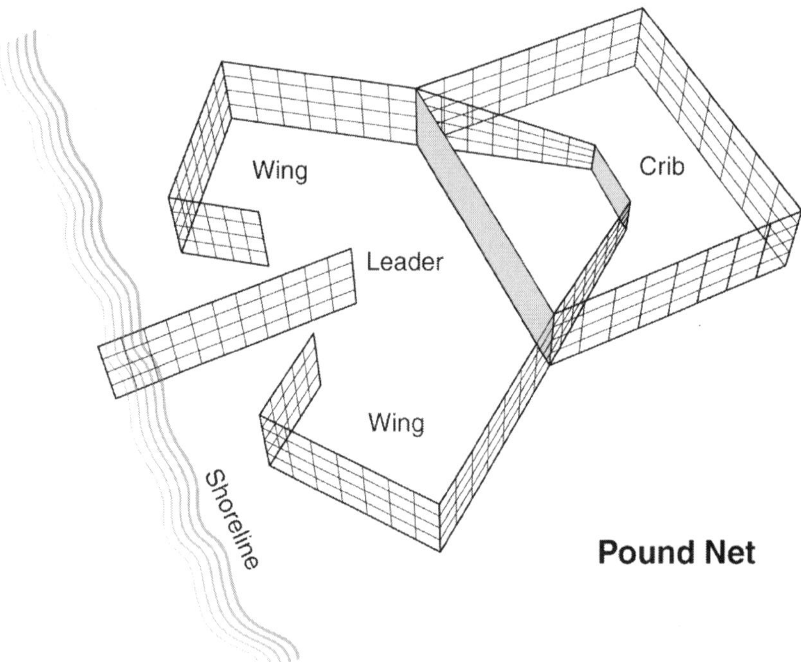

Fig. Typical configuration of a pound net used to intercept fish migrating along a shoreline.

Nets of one sort or another account for most of the fish caught in modern commercial fishing. Gill nets are passive devices that either capture the fish in their meshes or entangle them. When anchored into the bottom they may be used to catch ground fish. More commonly they are suspended vertically in the water column by a series of floats and weights. In the latter mode they are referred to as drift nets. The practice of using drift nets to catch fish has become highly controversial because the nets are simply too efficient. Unfortunately they often catch many fish and other organisms (*e.g.*, birds, marine mammals) that are not the target of the fishery. In the North Pacific, Japan, Taiwan, and South Korea for many years were engaged in a high seas drift net fishery that targeted squid, salmon, and billfish. Every night during the fishing season, a single vessel would set a 10-30 kilometer long drift net that was picked up the following morning.

Eisenbud has estimated that the length of all the nets used in the fishery totaled 33,000 km, roughly 80% of the circumference of the Earth. The use of high seas drift nets proved so destructive to both target and non-target species that in 1992 the United Nations General Assembly adopted a resolution requiring all nations involved in high seas drift net fishing to cease such operations by December 31 of that year. Nevertheless, drift nets continue to be used in many coastal fisheries.

The use of trawl nets (not to be confused with trawl lines) is an active form of fishing that involves pulling a net through the water or along the bottom to capture fish. An obviously limiting factor in the use of this technique is the location of the fish. If the target fish are demersal, then it makes sense to pull the net along the bottom.

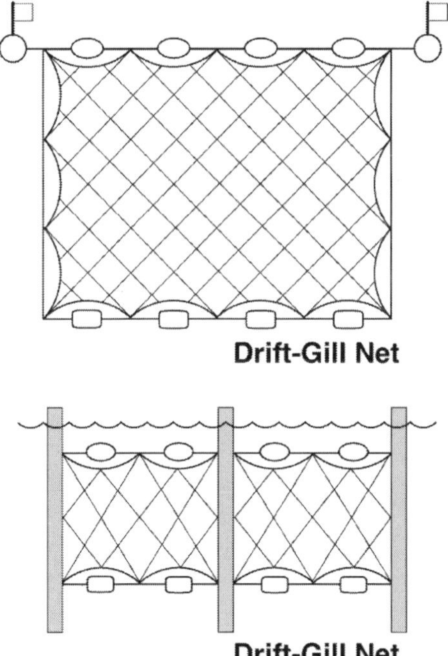

Fig. Illustration of drift gill net (upper panel) and stake gill net (lower panel).

However, dragging a net across the bottom is a difficult job, and the use of trawl nets to catch bottom fish did not become popular until steam and later fossil fuel powered vessels became standard fishing technology. Damage to nets was also a serious problem as long as the nets were made of natural fibers. Nets made of synthetic fibers proved far more durable for bottom trawling, but the technology is still impractical if the bottom is rugged or rocky. Even under ideal conditions, one of the early problems with the use of trawl nets was keeping the mouth of the net open as it was pulled through the water. This problem was initially solved by placing a beam across the mouth of the net, but this solution imposed a limit on the size of the net that could be used, since it is difficult to manage a large beam.

This problem was resolved by the development of the otter trawl, which was first used by the British during the latter half of the 19th century and later by Americans during the first half of the 20th century. The otter[1] trawl employs two broad, flat panels that are attached by lead lines called sweeps to opposite sides of the mouth of the trawl net.

The net is towed by lines called warps attached to these panels, and the force of water against the panels as the net is pulled through the water pushes the panels apart and keeps the mouth of the net open. This ingenious strategy facilitated the use of much larger nets than would have been possible with the beam trawl design. The mouth of a trawl net can range form 10 meters to as much as 50-120 meters across.

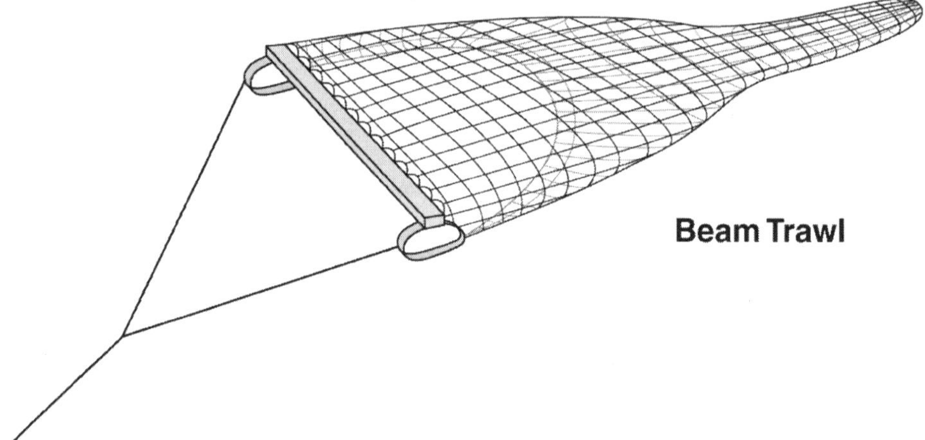

Fig. Illustration of a beam trawl.

Otter trawls are used to catch both bottom fish and pelagic fish. In the case of bottom fish, the targets include shrimp, cod, haddock, whiting, flounder, saithe, and fluke. Limiting factors are the depth and topography of the bottom. It is not very practical to go bottom trawling in a water column 5 km deep or if the bottom is covered with topographical features that do serious damage to the net.

In recent years an additional concern has been the impact of repeated bottom trawling on the benthic habitat. It has been argued, not illogically, that repeatedly dragging heavy nets over the seafloor does serious damage to the benthic habitat, not unlike the effects of clear cutting a forest. In the case of mid-water fish, an important limiting factor is the ability of the fisherman to know at what depth the net should be towed. Prior to the development of the echo-sounder, a mid-water trawl was essentially fishing in the dark. However, with the advent of this device, it became possible to locate schools of mid-water fish from a surface vessel, and by placing a depth recording device on the otter boards, the fisherman could control the depth of the tow and dramatically improve his chances of catching fish. Modern trawl nets are even more sophisticated.

They are equipped with sonar arrays that track the school of fish just ahead of the net. Bow sonar and computers automatically steer the ship towards the largest schools of fish. Fish targeted by mid-water trawls include cod, haddock, and hake, which sometimes aggregate in large schools far enough above the bottom to avoid capture by conventional bottom trawls. Trawl nets are now second only to purse seines with respect to the quantity of fish caught each year on a global basis, and in some parts of the world (*e.g.*, northwestern Europe) they account for more of the commercial fish catch than any other fishing method.

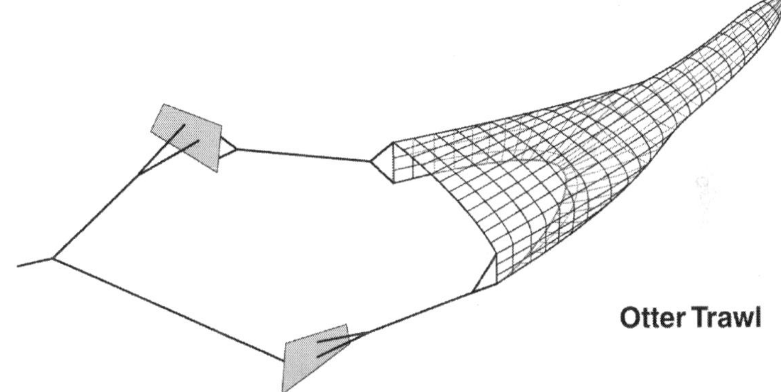

Fig. Illustration of an otter trawl.

Globally by far the most important technique for catching fish is the purse seine. Purse seines are used to catch fish that school near the surface. The device is a large net as much as 1.0 km long that extends perhaps 100 meters down into the water column. The strategy is to first encircle a school of fish with the net.

The bottom of the net is then closed by pursing the net, *i.e.*, by drawing in a rope that passes through a series of rings attached to the bottom of the net. This effectively seals off the bottom of the net, which is then pulled on board the boat until the fish are confined to a relatively small pocket or bunt from which they can be conveniently netted or (in the case of small fish) pumped into the hold of the fishing vessel.

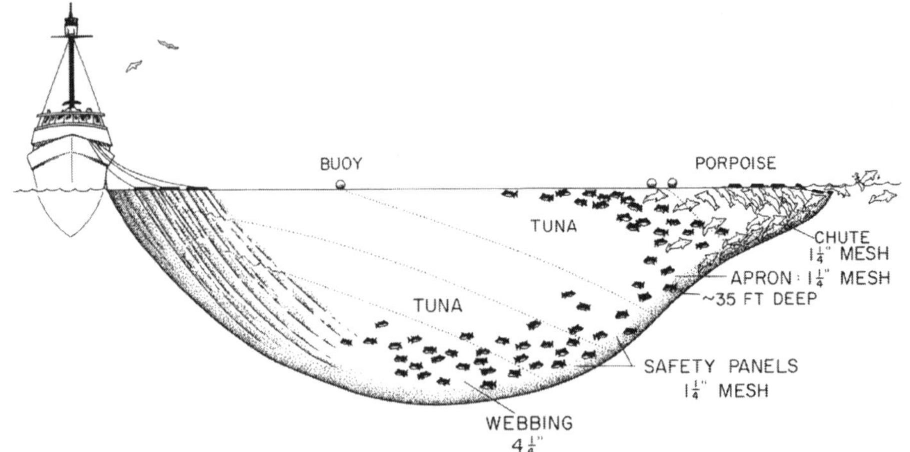

Fig. Illustration of a purse seine.

Several factors limit the use of this technology. First, the target fish must aggregate in large schools at or near the surface. Second, there must be a practical means for deploying and retrieving the very large nets used in purse seining. Prior to roughly 1950 it was common practice to set purse seine nets manually using small boats called dories. This technology (or lack thereof) limited the number of fish that could realistically be caught with a purse seine. This limitation was overcome with the use of the power block, which was invented by a California sardine and tuna fisherman named Mario Puretic. The device consists of a grooved pulley hung overboard from a boom and through which the entire net is drawn on board.

The pulley, or sheave, is driven by a compressed-air pump and has self-propelled rollers equipped with cleats that grip the net. Once the net passes through the pulley, it descends to the deck, where it neatly (more-or-less) folds down.

The third very practical problem is keeping the target fish in the net before it can be pursed. In the case of tuna, for example, the thermocline creates an effective barrier to escape. As long as the net descends into the thermocline, the tuna are disinclined to dive to freedom. The shallow thermoclines of eastern ocean basins thus facilitate tuna fishing with purse seines. Some commercially important species caught with purse seines include anchovies, menhaden, sardines, pilchards, herring, mackerel, tuna, and salmon.

COMMENTS AND OBSERVATIONS

With this introduction to fishing methods, it is instructive to summarize their quantitative importance to global capture fisheries. At the present time roughly 50% of the catch is taken by purse seine. The principal targeted species include the Peruvian anchovy (Anchoveta), Chilean jack mackerel, Atlantic herring, Japanese anchovy, skipjack tuna, capelin, chub mackerel, yellowfin tuna, and European pilchard. The major nations and fishing areas involved in the capture of these fish. Most of these species are caught near coastlines by a small number of countries in a well defined geographical area. Exceptions include Atlantic herring, which are taken on both sides of the North Atlantic by five major fishing nations, and skipjack and yellowfin tuna, which are caught throughout the western Pacific and Indian Ocean (and elsewhere) by numerous nations. Of the purse seine catch, about 50% of fish such as herring, sardines, pilchards, and anchovies are used for reduction purposes. Tuna and mackerel go primarily for human consumption.

Otter trawling accounts for about 17% of the catch, the most important species being Alaska pollock, largehead hairtail, Atlantic cod, and blue whiting. The first three are used primarily for human consumption. The majority of the blue whiting catch is reduced to fishmeal.

The major nations and fishing areas involved in otter trawl fishing. The list includes two species, Alaska Pollock and Atlantic cod, whose catch has declined significantly in recent years. During the 1960's the Atlantic cod catch averaged about 3 Mt y^{-1}. In the last few years the catch has been about 1 Mt y^{-1}, primarily due to overfishing in the northwestern Atlantic. The catch of Alaska Pollock has likewise declined from a peak of 5-6 Mt y^{-1} during the 1970's and 1980's to about 3 Mt y^{-1} during the first few years of the 21st century. Of the remaining commercial catch, about 9% is taken with fishing lines, 8% with pound and other trap nets, and 6% with gill nets.

The choice of gear used in a particular fishery reflects a variety of considerations. Among these are the horizontal and vertical distributions of the target fish, their tendency to aggregate at certain times of year, and water clarity. Fish that congregate in shoals or schools near the surface are likely to be fished with purse seines. Tows or drift gear are used when the target fish are more dispersed.

Otter trawls or trawl lines are used to catch ground fish. In the case of shrimp, fishing (with otter trawls) is done at night, since the shrimp burrow in during the day. Otter trawls or pelagic long lines are used to catch midwater fish. In the case of the otter trawls, echo sounders are crucial to successful use of the gear. The behaviour of some fish warrants use of different types of gear during different seasons or even at different times of day. Herring in the North Sea, for example, are fished with bottom trawls during the day and with pelagic gear at night.

Both cod and herring form dense aggregations while spawning in the North Sea. Knowledge of the location of the spawning grounds and timing of the spawning event has historically facilitated capture of these fish. The migrations of salmon from the ocean to their freshwater spawning grounds are one of the best and historically well documented examples of the way fishing has been facilitated by human knowledge of fish behaviour. Water clarity becomes a factor in fishing when the gear being used can be visually detected (and avoided) by the target fish. Gill nets, for example, are not very useful if they can be seen. It is for this reason that gill nets designed for use during daylight hours are made of synthetic material that is more-or-less invisible when hanging in the water. Historically, for example, purse seines were of little use for catching tuna around the Hawaiian Islands, because the water is clear and the thermocline deep. The clear water facilitated detection of the net by the fish, and the deep thermocline provided no barrier to escape before the net was pursed.

Finally, it is worth reiterating that technological developments have played a big part in the evolution of the commercial fishing industry. Echo sounders have been crucial to the location of schools of midwater fish. In the case of nets, synthetic fibers have dramatically changed the amount of time fishermen spend on the fishing grounds as opposed to mending their nets. Power, first in the form of steam and later in the form of fossil fuels, has had major impacts on the methodology and feasibility of fishing. Pulling bottom trawls and hauling in purse seines is now done with relative ease using nets that would have been utterly impossible to manipulate by hand. An important technological development in the case of otter trawling was the use of stern ramps for retrieving the net after the trawl. As recently as the end of World War II otter trawls were still being "shot"[2] from the side of the fishing vessel. At the end of the trawl, a boom was used to lift the cod end so that the catch could be discharged onto the deck or into the hold. The rationale for recovering the net from the side of the ship was concern over fouling the propeller if the net were brought on board from the stern. This problem was overcome by recessing the ship's screws. With this adjustment, it became possible to haul nets on board using a stern ramp, a development that permitted the use of much larger nets than had previously been possible. The idea of using stern ramps in this way undoubtedly derived from developments in the whaling industry, where stern ramps were routinely being used to haul aboard whale carcasses weighing as much as 100 tonnes.

The ability of boats to travel great distances to fishing grounds and to return safely with their catch directly reflects the availability of relatively cheap and reliable energy derived from fossil fuels to move the boats through the water. While long-distance fishing was by no means unheard of in the age of sailing, the prospects for fishing far from home are greatly enhanced when the fishing boat is powered with fossil fuel.

On such long fishing voyages, and indeed even when the ship returns to port after no more than a day or two at sea, preservation of the catch is a critical issue. Prior to the development of modern technology, standard methods for preserving the catch were drying or salting. In the former case the water content of the fish is reduced to a point where the multiplication of harmful bacteria is precluded. In the latter case the same effect is achieved by increasing the concentration of salt in the tissues of the fish. Historically fish were dried in the sun. The process works well with lean fish, but fatty fish such as herring, sardines, and anchovies are difficult to dry. The fat begins to go rancid before the water content is reduced sufficiently to block bacterial growth. The solution for fatty fish is salting, for which instructions can be found in cookbooks dating to the Roman Empire.

Drying and salting are still practiced in many parts of the world for cultural reasons and/or because more sophisticated methods of fish preservation are unavailable. Modern strategies for preserving fish involve either freezing or canning. The practice of canning is traceable to the early 19th century, when the French chef Nicolas Appert won a prize from his government for developing a new method of preserving foods that relied on heating under a seal. Initially the containers were made of glass. Preserving food in cans awaited the development of canning technology and advances in bacteriology that made canning a reliable method of food preservation. By 1900 canning fish had become a common practice, particularly with fatty fish such as salmon, sardines, anchovies, and later tuna. Preserving fish by freezing initially relied on the use of ice, which became widely available for such purposes in countries such as the United States during the latter half of the 19th century. To maintain the quality of the product, the fish, once frozen, must be maintained in a frozen condition until it is thawed for cooking. A major advancement in freezing technology was the development of the plate freezer by Clarence Birdseye in 1929. With ice the temperature could at best be reduced to 0°C. With the freezing technology developed by Birdseye it was possible to deep freeze fish and thus ensure preservation from initial capture to the time the fish is thawed for cooking. The fish that freeze best are lean fish that do not can well such as cod and other so-called white fish.

This review of commercial fishing methods and technology has focused for the most part on large-scale company-owned commercial fishing. It is worthwhile to point out that substantial numbers of fish are still caught by small-scale artisanal fishermen using methods that are much less sophisticated and expensive than those employed by modern fishing boats. The provocative comparison of the two fishing industries.

Although the figure is based on an article published in 1980, the basic characteristics of the two fishing industries are accurately captured in this comparison. Far more persons are employed in the small-scale artisanal fishing industry, and per dollar invested in fishing boats, the small-scale industry is a much more effective way of keeping people employed. Virtually the entire catch from the small-scale industry goes for human consumption. Roughly 45% of the catch from the large-scale company-owned industry is used for reduction purposes.

For several decades there has been a real concern that the large-scale company-owned fishing industry would eventually eliminate most artianal fishermen. While the balance of competition favored the former during much of the 20th century, increases in the cost of fossil fuels may eventually shift the advantage to the small-scale industry. It is noteworthy that artisanal fishermen consume 5-10 times less fuel per tonne of fish caught than their counterparts in the large-scale company-owned industry. Many of the technological developments that company-owned fishing boats have been able to exploit to find and catch fish and bring them to market have made sense financially because of the low cost of fossil fuels. It is provocative to imagine how the commercial fishing industry would be structured if the price of oil were $100 per barrel, a scenario many persons now anticipate to occur sometime during the first half of the 21st century.

IMPACT OF ARTIFICIAL REEFS

ENHANCEMENT AND RECRUITMENT

It takes weeks for a new artificial reef to attract fish and for the establishment of fish colonies/populations. Artificial reefs mainly concentrate on the remaining resource and the potential enhancement effect is, in general, considered quite low. In terms of habitat rehabilitation, artificial reefs have little, if any, success as they only concern a limited area.

It is worth observing that the assessment of some possible habitat rehabilitation and production enhancement would require a careful survey over a long period with regular data collection and careful monitoring of all fishing activities and in-depth knowledge of local ecosystems and ecological mechanisms. In most cases it is very difficult to evaluate fishing activity on artificial reefs. Therefore, regarding the potential socio-economic impact, there is a lack of thorough economic analyses of the costs and benefits of artificial reefs.

In general, the establishment of artificial reefs in coastal areas can facilitate smaller scale fishing activities with passive gears such as lines, pots and gillnets. In some places, recreational fishers also benefit from the creation of artificial reefs.

BIOLOGICAL

Regarding the biological impact of artificial reefs, two theories exist concerning the number of fish and fish species using reefs:

- Competition for limited resources, such as food and shelter is the chief factor; and,
- Fish mortality and recruitment are essential. The potential advantage from reefs could be a reduction in mortality (for instance by providing

refuge from predators for small fish/juveniles) and/or recruitment stability.

Fish behaviour on artificial reefs differs from one species to another. This fact determines the different aggregating ability of the artificial reef for various species.

LEGAL ASPECTS AND REGULATION

Legal aspects related to artificial reefs include:

- Ownership of the artificial reef;
- Ownership of the fish on it;
- Limitation/prevention of certain fishing gear and therefore certain fisheries, resource utilization and management; and,
- Shipping and navigation.

A number of countries have set up strict regulations concerning the position of the artificial reefs, design and material used, proper marking, issue of a government authorization, amount of fishing gear authorized, regular reporting of fishing activities on artificial reefs, etc.

Any national regulation must take into consideration the provisions of UNCLOS (United Nations Convention on the Law of the Sea), conventions on navigational safety, shipping routes, conventions on dumping at sea, and decisions by fisheries management bodies.

When offshore structures for oil or gas exploitation act as a permanent fish aggregating device or artificial reef, they are covered by international standards and guidelines developed by the International Maritime Organization (IMO).

The creation of artificial reefs has changed the perception of fishery resources from a primarily competitive hunting attitude to a collaborative rearing and nursing one.

In addition to direct impact of local fisheries, performances and revenues, artificial reefs have side-impacts on other fisheries and related activities in terms of catches and incomes.

Where there is heavy or overexploitation, deployment of artificial reefs appears to become an alternative to more traditional development efforts of artisanal fisheries for effective near-shore fishery resources management. It is worth mentioning that the settlement of such structures introduces important changes in the management of fishing operations: *e.g.* fishing rights, working time allocation, relationship among fisher groups and between small and larger scale fisheries.

AQUACULTURE TECHNOLOGY

The technology employed in aquaculture has developed over many centuries but has done so more rapidly during the last half century. Aquaculture systems, and the technology used, vary from very simple systems, used for family ponds in tropical countries, where production is for domestic consumption, to high technology systems, such as intensive closed systems, like those used for rearing stripped bass. Herbivorous and filter feeding fish, reared mostly in simple systems of small freshwater ponds, however, account for about half of global aquaculture production.

Much of the technology used in aquaculture is relatively simple, amounting to small modifications that improve the growth and survival rates of the target species, such as providing additional food, adding seed animals collected elsewhere, managing water exchange to maintain adequate oxygen levels, and protecting the stock from predators. Greater understanding of complex interactions between nutrients, bacteria and cultured organisms, together with advances in hydrodynamics applied to pond and tank design, have enabled the development of closed systems. These have the advantage of isolating the aquaculture systems from natural aquatic systems, thus minimizing the risk of disease or genetic impacts on the external systems.

Developments in engineering, some learnt from offshore oilrig construction, increase the possibilities for offshore aquaculture using robust cages. Sea ranching, the release of young fish into the wild to improve the harvest in capture fisheries, has also made a start but its long term viability is still to be assessed. Major advances are also being made in the technology of the production of aquafeeds, which generally require the combining of a large number of ingredients into very small feed pellets.

Aquaculture Engineering

Aquaculture draws on well-established engineering fields for most of the design and construction needs of its production facilities. Building earth ponds is similar to building roads - a knowledge of the characteristics of soils and the limits of safe design are the basis of good construction. Similarly, the buildings used for hatcheries and other support activities are no different from those common in the housing, agricultural and commercial sectors. Sometimes ponds are lined with plastics or other impermeable materials, and here the techniques are similar to those for civil structures such as potable water reservoirs or sludge tanks. The design and installation of water control gates, including in unstable soils, benefits from the long experience in this field in the agriculture and irrigation sectors. An understanding of hydrodynamics allows ponds and tanks to be built with good water circulation, oxygen mixing and without 'dead spots' where sediments might accumulate and cause health problems to fish.

For installations in the sea, the situation is somewhat different and many of the important engineering solutions, such as for fish cages or suspended shellfish growout systems, have had to be developed by aquaculturists themselves. They have benefited however from the accumulated knowledge of seafarers in general and fishermen in particular, in the design and operation of mooring and buoyage systems. More recently, when fish farmers have turned their attention to how to operate fish cages in locations further offshore where seas are rougher, the experience of the oil exploration industry has proven very valuable. Techniques have been developed in recent years for the production of fish and other aquatic products in closed recirculation systems. To make these work, aquaculturists have needed to develop a knowledge of the biological processes operating - such as how bacteria can be used to neutralise and re-cycle the nitrogenous waste products produced by growing fish - and how to engineer the systems to meet the biological requirements. Knowledge of bio-engineering from the waste treatment and water treatment industries has made contributions to the development of closed aquaculture systems and there is probably more that could be usefully transferred from the sewage treatment sector to help solve problems in fish rearing.

Modern materials have brought many benefits to aquaculture. For instance, custom made plastic joints have simplified the construction of sea cages, and made them more reliable in high stress conditions. Experiments with huge free-floating or sunken net cages operated in the open ocean were begun several decades ago, for instance in the Caspian Sea. These showed some promise, but more reliable construction materials will make the farming of fish in such structures increasingly feasible. Modern materials and production methods have been important also, in the construction of plastic filter substrates for indoor recirculating systems. The fine detail of these has been found to make substantial differences to the efficiency of biological filters.In shrimp farms, specially designed matting materials that stand upright on pond bottoms, with a structure that promotes the growth of the small animals and plants that the shrimp can thrive on, have recently been developed and shown to boost production.

Engineering skills are important in the design of most aquaculture facilities and good engineering can affect the efficiency and economics of production. If capital costs can be minimised while still maximising productivity and reducing risk, the farming operation will be more profitable. Aquaculturists have proven very innovative over the past fifty years, constantly developing new technologies to support their farming operations. As new production methods and species for farming are developed, the engineering solutions needed to support them will continue to evolve.

Aquaculture Facilities

Aquaculture began with man making small modifications to natural habitats so as to improve the survival and growth of target species. Some of the oldest examples are in the rearing of freshwater fish in ponds, which has been practiced for thousands of years in Asia and at least for many centuries in Europe. The simple act of placing a mesh barrier across the outlet of a small pond or lake to prevent fish from escaping can make a big improvement in the supply of food. Similarly, along coasts, some tidal lagoons can be easily turned into ponds. Closing off such naturally occurring water bodies was the start, centuries ago, of much fish and shrimp aquaculture in Asia and, in modern times, in South America. Removing predators and improving the conditions within the pond, (for instance by providing more area of preferred water depth), supplying additional food and, later, by adding seed animals collected outside, were further steps that moved aquaculture production close to where it is today. The farming of seaweeds and molluscs (oysters, clams, mussels etc) developed similarly, as people made improvements such as providing more settlement areas for the young, or.the removal of predators from the growing area.

Farmers had a natural desire to improve the productivity of their systems and, as knowledge grew, they learned to stock more animals, increase feeding and manage the exchange of water to maintain the conditions, such as adequate oxygen levels, that the animals needed to survive. The range of facilities used for aquaculture subsequently broadened for a number of reasons. Farmers encountered difficulties with more intensive rearing because of the uncontrolled influence of pond soils, local water quality, weather. Some of these problems could be resolved by rearing in ponds built of concrete, or lined with plastic, by bringing the ponds indoors under cover, or treating the water before flowing it to the culture ponds.

Secondly, as the naturally occurring ponds and lagoons all became used, prospective farmers had to take a broader approach and develop the technology and engineering to be able to use less naturally-favored sites. Net cages floating in protected coastal or inland waters were developed for fish culture, or fish were stocked in fenced areas of the sea or of large lakes. At the same time, the need for supplies of young animals (fry, seed) to stock these systems led to the development of hatchery techniques and dedicated hatchery facilities. For most species this aspect of production has proved more successful when conditions can be more closely controlled, for instance in indoor concrete and fibreglass tank systems, rather than in outdoor ponds. Often, the younger stages of aquatic animals are more sensitive than adults to physical and chemical conditions and these have to be managed within a smaller range, if production is to be successful. Thus hatchery facilities developed as a separate branch of the industry.

More recently, our knowledge has improved greatly regarding the complex interactions that occur in a rearing system, between nutrients, bacteria and the cultured organism. This and technological developments have allowed many aquatic organisms to be reared in completely closed recirculating facilities, including the farming of marine organisms at locations far from the sea. Closed systems have the added advantage of offering greater protection from the danger of disease entering from the natural environment and also of minimizing adverse effects of the production system on that external environment. Technology has also begun to open up the possibilities of growing fish in enclosures in the open ocean, something that could one day transform the nature of human food production on the planet. With 70% of the earth's surface covered by water, the potential is clear. Earlier technology has restricted the cage farming of fish to sheltered coastal waters. Here the number of available sites is limited, environmental damage is more likely and conflicts exist with other users. As cages are developed that can withstand the demanding conditions of the open ocean, or which can be operated below the ocean surface, farming could move further offshore.

A small start has been made, notably in Japan, through so-called 'ranching' programmes', the farming of the sea without enclosures where fry are released in large numbers into the ocean with the aim of improving returns from the capture fishery. Sometimes artificial reef structures are created underwater as well, to increase the available habitat and natural food for the fish.

Aquaculture Systems

Aquaculture - the growing of aquatic animals and plants - covers a wide range of species and methods. Some of the simplest production systems are the small family ponds in tropical countries where carp are reared for domestic consumption. At the other end of the scale are high technology systems, such as the intensive indoor closed units used in the USA for the rearing of striped bass or the sea cages used in Chile and Europe for growing salmon and bream. Nearly half of the world's aquaculture however is of herbivorous and filter-feeding fish such as carp.

Practised in freshwater, in ponds made of earth, and often managed by a single household, production systems are simple and in many cases have changed little over many centuries. A farmer will excavate a small pond near his house; sometimes, as for instance in the flood river deltas of countries like Vietnam, using the spoil to build a raised foundation for the house itself. The pond will typically be fed by natural rainfall or groundwater, or sometimes by diverting a nearby stream or irrigation canal. Some stocking occurs naturally, but more commonly a farmer will buy young fish from a breeder and stock them in his pond, often after holding them for a time in a small floating net known as a 'hapa' to check that they are in good condition before release in the pond. Production will improve if the farmer can stock a good balance of different species to make the best use of the varied kinds of food available in the pond. Typically the fish will be fed with household or agricultural by-products, and harvested when the family needs a meal.

Stocking and harvesting is often continuous, where knowledge of the best stocking level is learned from experience and where the pond is not drained for many years. For most such farmers, fish production is a secondary activity, a useful additional source of protein to add to the supply or income from his main agricultural or commercial activities.

Outdoor freshwater fish farming is practised commercially in many countries, including in the developed nations. The density of stocking may be much higher, the control of feeding, water quality and fish health more closely monitored, but even in intensively operated systems, the key parameters that need to be monitored and balanced for success are similar to those in the traditional household pond system.

Over the last half-century, systems for the indoor rearing of many species of fish have also been developed. Holding of fish in controlled conditions indoors has been important in the development of seed production (hatcheries). Also, as knowledge has increased of nutrient cycles, bacterial action and water chemistry, it has become possible to rear fish, both for food and for ornament, in 'closed' indoor systems, where the water is recirculated. Passing the water through filters that use bacteria to naturally break down and recycle the waste products produced by the fish allows a production system to be run in almost total isolation from the outside. Such systems have been important in the control of disease and also in maintaining the stable conditions that some species need to flourish and reproduce.

A fifth of current world aquaculture production is of plants - mainly seaweeds, that for the most part are grown on the seabed or on raft or racks in shallow coastal waters and used directly for food or for the production of alginate or carageenan (agar-agar). This sector of aquaculture is another largely run by small-scale growers, mainly in Asia but in some parts of the South America, who attach seedlings of marine plants to simple structures built close to shore and constructed from natural materials and then harvest the plants once they have grown. A further fifth of world production is of molluscs such as oysters, clams and mussels.

Again the majority of production is small-scale and by coastal people who often combine fishing activities with farming. Oysters and mussels are mainly grown on structures built above the seabed - poles or racks on the shore, or ropes suspended from rafts or floating lines. The farmer's role is to supply suitable places for seed to settle or to add to natural production by bringing seed from a hatchery - and then to maintain the conditions of waterflow and freedom from predators that the shellfish need to grow. Most commercially grown molluscs feed on microscopic algae floating in the water, so the farmer does not need to provide any feed.

Fish are also grown in coastal areas, in ponds or in floating cages. As technology advances, there is the potential to develop systems for rearing fish in the open ocean, either in sturdy cages or by so called 'ranching', where young fish are released to the wild and then collected by normal fishing or by training them to respond to specially generated sounds. The tropical coastal zones of Asia and Latin America are where most shrimp farms are found. Shrimp farming is mainly carried out in earth-bottomed ponds built on flat land close to the sea. Shrimp are raised in ponds at a range of different densities. At low densities the systems need only limited inputs of feed and fertilizers. As the density increases, more feed has to be supplied and at the higher end of the range, so-called 'intensive' farming, machines have to be put in the ponds to mix and aerate the water.

Aquaculture is practised from the cold waters of the far north and south, where fish like salmon, arctic char and sturgeon are grown in ponds, flowing raceways and cages in the sea, down through the latitudes to the tropics, where carp and tilapia flourish in freshwater and shrimp and seabass are farmed along the coasts. It ranges from production of fish in naturally occurring ponds in rural areas to the intensive culture of ornamental fish in plastic tanks in the middle of a city. It is practised by the poorest farmers in developing countries as a livelihood and to supply much needed protein for their families - and by urban sports shop owners in Europe and the U.S.A. producing baitfish for weekend anglers. Systems can range from an intensive indoor system monitored with high-tech equipment through to the simple release of baby fish to the sea - but all with the same aim, of helping boost Nature's natural productivity

Biotechnology Development

Biotechnology - any technological application that uses biological systems, living organisms, or derivatives thereof, to make or modify products or processes for specific use 1994 UNEP Convention on Biological Diversity - has a wide range of useful applications in fisheries and aquaculture. It creates opportunities, for instance, to increase growth rate in farmed species, boost the nutritional value of aquafeeds, improve fish health, help restore and protect environments, extend the range of aquatic species and improve the management and conservation of wild stocks.

Some biotechnologies are simple with a long history of application such as the fertilization of ponds to increase feed availability. Others are more advanced and take advantage of increasing knowledge of molecular biology and genetics, *e.g.* genetic engineering and DNA disease diagnosis.The field of genetic biotechnology similarly ranges from simple techniques such as hybridization, to more complex processes such as the transfer of specific genes between species to create GMOs (genetically modified organisms).

Over the years, our knowledge of fish breeding requirements has improved and the ability to induce artificial breeding developed through the use of natural or synthetic hormones and/or environmental manipulations. (For example changing photoperiod or water temperature can induce some fish to spawn). These have been key factors facilitating the application of more advanced biotechnologies. Selective breeding, the maintenance of stocks genetically improved by chromosome manipulation, line crossing, and sex reversal all depend on the controlled breeding of farmed species. These improvements in reproductive technologies have also assisted aquaculturists greatly in their efforts to domesticate aquatic animals.

In addition, by making it possible to remove the natural constraints and timing of breeding, farmers are able to mate many more species at times that are most beneficial, and thus helping to ensure a steady and consistent supply of fish for consumption.

FEEDS TECHNOLOGY IN AQUACULTURE

Although aquaculture dates from the earliest parts of human history in Asia, Europe and in the Pacific Islands, it is only in the last few decades that the sector has begun to catch up with the rest of animal agriculture in terms of the science of feed milling and nutrition. Despite this, aquaculture currently represents the fastest growing segment of agriculture and the animal feed milling industry, particularly in China and the Asian region where over 90% of global aquaculture production is currently realized.

A very wide range of ingredients are used to prepare aquafeeds. Feeds range from single component feeds available on-farm such as grass or rice bran to farm-made formulated feeds and commercial feeds. They include aquatic and terrestrial plants (duckweeds, azolla, water hyacinth etc.), aquatic animals (snails, clams etc.) and terrestrial-based live feeds (silkworm larvae, maggots etc.), plant processing products (de-oiled cakes and meals, beans, grains and brans) and animal-processing by-products (blood and feather meal, bone meal etc.).

Formulated commercial feeds are composed of several ingredients, mixed in various proportions to complement each other and form a nutritionally complete diet. Farm-made and commercial aquafeeds can be fairly easily split depending on whether they are primarily intended for farm use or for commercial sale. Also, while raw materials that are high in moisture and only of local and/or seasonal availability may be used in the preparation of farm-made aquafeeds, commercial fish feed manufacture is predominantly associated with the processing of dry ingredients and the manufacture of a dry product.

The most common aquafeed processing operations can be summarized as raw material size reduction, raw material blending, feed forming, and feed drying. For these processes many and various options of processing equipment are available ranging from simple mortar and pestle to mincers, hammer mills, pelleters, and extruders. Commercial aquafeed manufacture presents special challenges to the traditional feed milling concepts due to the size and variety of animals being cultivated. Moreover, feed for aquatic species requires a higher degree of precision be it the generally finer particle size of ingredients, or the precise mixing of as many as four dozen ingredients into a feed pellet of minute size in comparison to its terrestrial counterpart. These are the compelling reasons why many new aquaculture feed mills are dedicated to produce only aquatic feeds and often employ human food standards in production. Along with the higher standards of production come more expensive and higher quality standards for the ingredients used for what are often very sensitive production animals.

Farm-made Aquafeeds

There is a lack of an exact definition for farm-made aquafeeds. FAO have suggested that farm-made feeds be defined as feeds in pellet or other forms, consisting of one or more artificial and/or natural feedstuffs, produced for the exclusive use of a particular farming activity and not for commercial sale or profit.

FISHING GEARS

When studying escapes from trawl codends and other selective devices, the most common method of capturing or 'sampling' escaping fish is using a codend cover. This chapter will summarise a programme of work providing evidence for two potential problems with codend covers, when sampling fish for use in survival experiments.

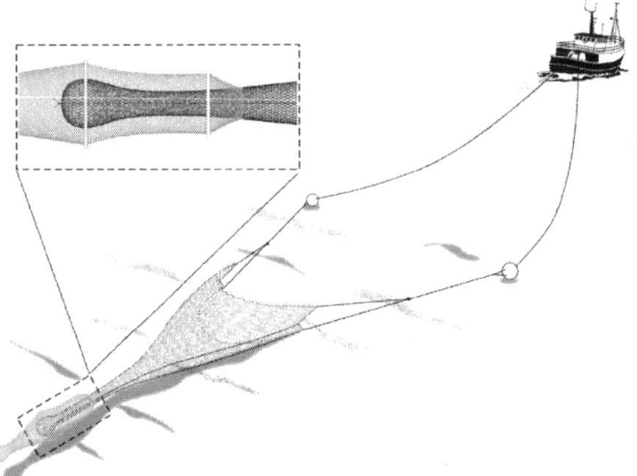

Fig. Trawl with Codend Cover attached.

DIRECT OBSERVATIONS

A series of direct observations, using divers and remote underwater cameras, were made of the behaviour of gadoid fish (in particular haddock) escaping from a trawl codend into a 'survival' cover during a research cruise aboard FRV Clupea. The 'survival' cover was the same as that used in escape survival experiments undertaken by FRS Marine Laboratory and partners. The cover had originally been designed with the protection of the captive sample in mind. It was constructed of a small, square mesh (30mm), knot-less, nylon netting which was thought to minimise abrasive contact with the captive fish and restrict water-flow within the cover.

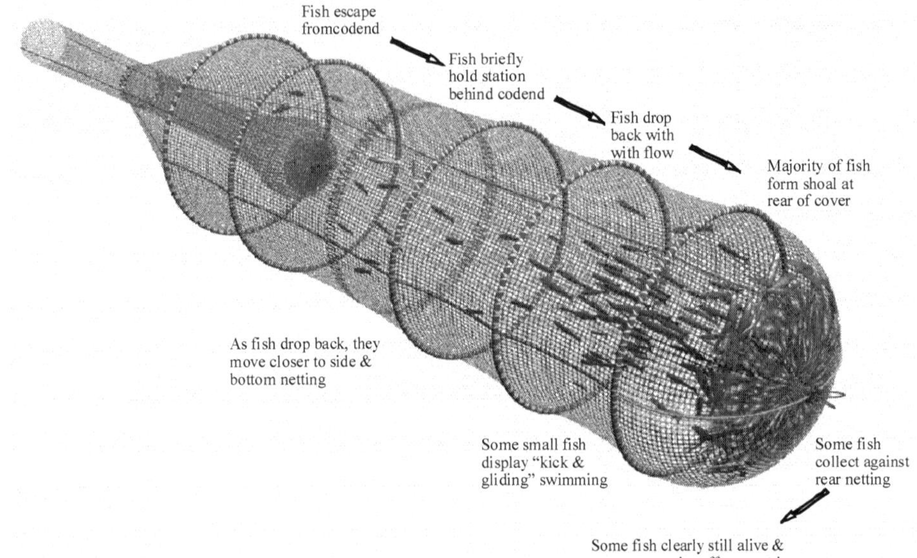

Fig..: Behaviour of Captive Fish in Codend Cover

The observed behaviour of the fish captive within this cover is summarise. The most relevant observation was that a significant number of fish were seen collecting on the netting at the rear of the cover, after only a few minutes of sampling time. It had been assumed that these fish had been killed or were close to death following their passage through the trawl and codend. However, closer inspection by divers revealed that many of these fish were in-fact alive, some repeatedly attempting to swim off the netting, but were trapped there by a substantial flow of water. What was unclear was whether or not these fish were trapped because of injuries or exhaustion received as a result of their passage through the trawl and codend.

FLOW MEASUREMENT

Detailed measurements were made of the water-flow (relative to the cover and codend) within the 'survival' cover both on half scale models, in a flume tank, and on the full scale cover at sea. It was revealed that while flow within the main body of the cover was indeed reduced, at operational towing speeds a substantial flow of water was still present (~0.45 m/sec for a trawl towing at ~1.4 m/sec). The likely impact of the water-flow upon the captive sample of fish was unknown.

In addition, it was shown that the water flow around the covered codend was significantly reduced (by ~80%) in comparison to that observed around a normal, uncovered codend. This has important implications for the behaviour of escaping fish and their likelihood of receiving an injury during their passage through the codend meshes. Firstly, the reduced flow will mean the codend experiences less drag and therefore the twine forming the codend meshes will be under less tension.

Passage through these meshes by escaping fish will, as a result, be easier and less injurious. Secondly, the passage from within the codend to the fast flowing water mass outside (relative to the codend and escaping fish) would be less hazardous because of reduced cross-flows. Again resulting in a less injurious exit from the codend.

In essence, the presence of the 'survival' cover around the trawl codend will protect escaping fish from injurious forces normally experienced during passage through a trawl codend.

This would suggest that this method of sampling escaping fish may underestimate the true escape mortality.

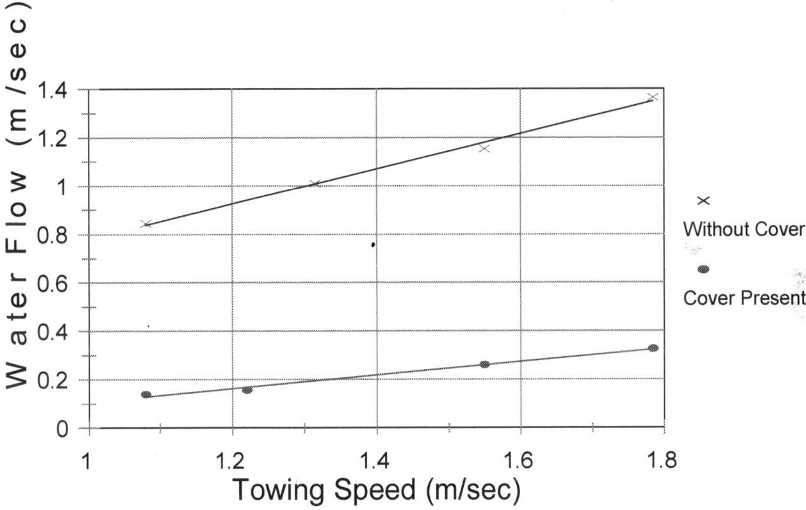

Fig. Water Flow Adjacent to Escape Zone on Trawl Codent

Swimming Experiments

An experiment was undertaken to investigate the prolonged swimming performance of haddock and estimate maximum sustainable swimming speeds. The method used to measure swimming endurance is based on the technique pioneered by He & Wardle using a rotating gantry with light array to stimulate the fish, via the optomotor response, to swim around a circular swimming path in the annular tank at the Marine Laboratory, Aberdeen. Individual haddock and groups of haddock were entrained to swim at a range of speeds (0.5 – 0.8m/sec) at a constant temperature (10 ± 0.5°C).

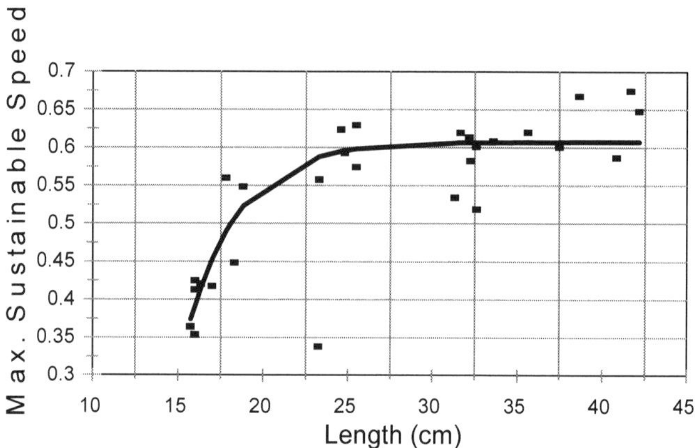

Fig. Max Sustainable Swimming Speed vs. Length.

Endurance was shown to be highly dependent upon the swimming speed and size of individual fish. Maximum sustainable swimming speeds were estimated for individual fish from their endurance data and shown to be related to length, particularly among smaller fish. Comparison of this relationship with the water flows experienced in the main body of the cover at operational speeds (~0.45 m/sec), suggests that the majority of smaller haddock (<18cm) would be unable to sustain the swimming speeds necessary to maintain position within the cover. These fish would be forced against the rear of the cover were they may receive a number of life threatening injuries, including skin abrasion and suffocation.

Reviewing Historic Data

The evidence from the swimming performance trials would suggest that water flow conditions within the 'survival' cover have induced a significant biasing mortality, particularly among smaller fish, in previous escape mortality investigations. If this were true, examination of past work should reveal two distinct trends. Firstly, it would be expected that escape mortality in haddock would be inversely proportional to length. This has been demonstrated in three separate experiments. Secondly, mortality would be expected to increase with increasing sampling time (ie. the period of time from when fish start to be collected in the cover to when it is released from the trawl). This relationship was confirmed by Breen et al when data from previous FRS escape survival investigations was modelled with respect to 15 potential explanatory variables, of which sampling time and mesh size proved to be the most significant factors.

SAMPLING TIME AND MESH SIZE FOR HADDOCK

For haddock at least, current designs of codend cover are inappropriate for sampling fish escaping from towed fishing gears for use in survival experiments.

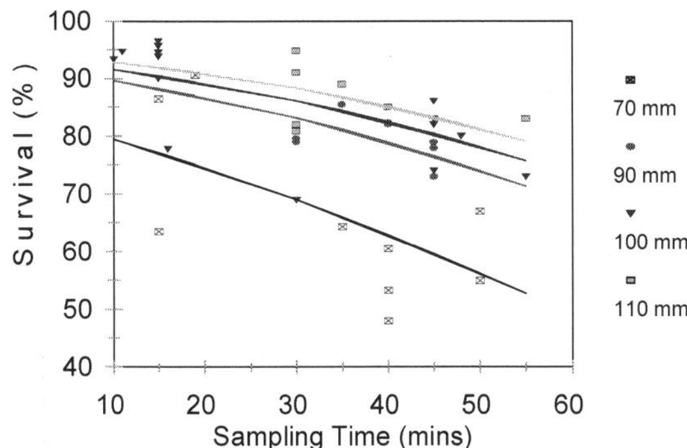

Fig. Relationship between Escape Survival, Sampling Time and Mesh Size for Haddock

Two confounding biases have been highlighted. Firstly, the presence of a cover significantly reduces the water flow around the codend (by ~80%), which is likely to affect escape behaviour and reduce the likelihood of injury during escape. Secondly, while the flow around the codend is reduced, there is still a substantial flow of water in the main body of the cover (~0.45 m/sec for a trawl towing at ~1.4 m/sec). From swimming endurance experiments on haddock, it is clear that the majority of smaller fish (<18cm) captive in the cover would not be able to sustain the swimming speeds demanded by these observed water flows. These fish are likely to receive life threatening injuries as a result of their captivity in the cover.

2

Classification of Fishes

There are about 28,100 species of fishes known to science, they are divided into 4 classes, 59 orders, 490 families and 4,300 or so genera. The classification system of fish is not stagnant and it will change when we have more knowledge, perhaps from DNA analyses.

INTRODUCTION

There are a few problems with the classification of fish, the first of which is the question 'What is a fish'? We all use the term Fish regularly but the term fish is not a natural group of animals in the way the terms Reptiles or Birds are. When scientists arrange living things into groups for classification they try to make the classification system they use represent the actual relationships between those living things. The more categories that two organisms share the more closely they are related.

All living things are divided up into a series of classifications delineated as:- Kingdom, Phylum, Class, Order, Family, Genus and species. You can also have Super and Sub attached to each group so that a Superclass can contain several Classes each of which itself contains several Subclasses, Subclasses then contain Superorders which contain Orders etc. So you see we can end up with a lot of different terms, which is good because there are a lot of living things.

The term 'Fish' does not fit into any of these categories. This is because it is a general term not a scientific one. We all know that a Trout or a Salmon is a fish, however some of the animals we commonly call fish are not much more closely related to Salmon and Trout than they are to Dogs and Cats. The term fish is actually used to describe any animal that is part of the Subphylum Vertebrata but is not a member of the Classes Amphibia, Reptilia, Aves (Birds) or Mammalia. The Subphylum Vertebrata contains two Superclasses, Agnatha and Gnathostomata. The Superclass Agnatha contains only two Classes of animals both of which come under the general term of fish. The superclass Gnathostomata contains six Classes of animals two of which are also included in the general term of fish.

Table: Higher Classification of the Phylum Chordata

Phylum	Subphylum	Superclass	Class	Common Name
Chordata	Urochordata	—	—	Tunicates
Chordata	Cephalochordata	—	—	Lancelets
Chordata	Vertebrata	Agnatha	Myxini	Hagfish (Fish)
Chordata	Vertebrata	Agnatha	Cephalspidomorphi	Lampreys (Fish)
Chordata	Vertebrata	Gnathostomata	Chondrichthyes	Sharks and Rays etc. (Fish)
Chordata	Vertebrata	Gnathostomata	Osteichthyes	Trout and Salmon etc. (Fish)
Chordata	Vertebrata	Gnathostomata	Amphibia	Amphibians
Chordata	Vertebrata	Gnathostomata	Reptilia	Snakes, Turtles etc.
Chordata	Vertebrata	Gnathostomata	Aves	Birds
Chordata	Vertebrata	Gnathostomata	Mammalia	Cats, Cows and People

The animals we commonly refer to as fish then can be seen to be included in 4 classes of animals that are as different from each other as Birds are from Reptiles, but because they all live in the sea they get lumped together.

When talking about, and reading about species of fish when only a common name is given you need to be careful. Many people know that Bonito is a type of tuna. We find that all the following 20 species are all called Bonito somewhere in the world. *Abramites eques, Auxis rochei rochei, Auxis thazard thazard, Caranx crysos, Cybiosarda elegans, Euthynnus affinis, Euthynnus alletteratus, Euthynnus lineatus, Gymnosarda unicolo, Katsuwonus pelamis, Megalaspis cordyla, Orcynopsis unicolor, Rachycentron canadum, Sarda australis, Sarda chiliensis chiliensis, Sarda chiliensis lineolata, Sarda orientalis, Sarda sarda, Selar boops, Thunnus alalunga.*

This is not just a list of the different species called bonito in different parts of the world. In many countries more than one species is known as Bonito, in fact in England, Australia and Cuba the term Bonito applies to 3 different species, in Kenya, Namibia, Brazil and Germany is applies to 4 different species, in Colombia and Mexico Bonito includes 5 species and in the USA the name Bonito is applied to 10 different species.

When you add to this the fact that the same species is also called different names in different countries the problem of knowing exactly what species a common name refers to can get quite difficult. For Instance:- *Katsuwonus pelamis* is Striped bonito, Ocean Bonito, Oceanic Bonito, Arctic Bonito and Bonito in the USA, Striped-bellied Bonito, Striped Bonito, Oceanic Bonito, Atlantic Bonito and Bonito in England, Bauchstreifiger Bonito in Germany, Bonitol, Bonito de altura, Bonito de veintre rayado and Bonito del sur in Spain, Bonito, Bonito de barriga listada, Bonito-de-barriga-riscada, Bonito de veintre rayado, Bonito-oceânico and Bonito rajado in Brazil.

METHODS AND DIFFERENCES BETWEEN CLASSIFICATIONS

Most ichthyologists adopt a cladistic methodology of classification, which classifies strictly according to recency of common descent. In this system, the goal is to have each taxonomic unit contain all species sharing a common ancestor. This often results in a very different classification than would result from the approach commonly used up to about 1980. In the older approach, evolutionary or synthetic systematics, the inferred amount of divergence among groups was considered, as was the history of their evolutionary relationships, and a given taxon might not contain all the descendants from the inferred ancestral species.

An example of the difference between these 2 approaches comes from showing hagfishes and lampreys in a classification. In a cladistic system, HAGFISHES would be placed in a category separate from LAMPREYS and other vertebrates because, relative to hagfishes, lampreys are thought by most ichthyologists using morphological evidence to share a common genealogical branching point with all other vertebrates. Systematists using the evolutionary approach could express the relationship in several ways but might prefer to place hagfishes and lampreys together in the category of jawless fishes, separate from all jawed vertebrates, although recognizing that they are very distantly related.

Differences may also exist in classifications because of ichthyologists using different characters, for example, molecular versus morphological characters. Other differences may also exist among classification systems, depending on whether the worker is a "lumper" or a "splitter." A lumper constructs a classification with categories containing more groups than that of a splitter. There are no rules for the amount of difference necessary to designate various categories. A lumper recognizes one family for SALMON, WHITEFISH and GRAYLING; a splitter, 2 or 3 families.

Classification

The following summarizes the major groups of fishes. Fishes and the tetrapods belong to the subphylum Craniata.

Superclass Myxinomorphi

Class Myxini

The remaining fishes and the tetrapods belong to the infraphylum Vertebrata.

Superclass Petromyzontomorphi

Class Petromyzontida

Superclass Gnathostomata - Jawed Vertebrates

Class Placodermi (extinct)

Class Chondrichthyes

Class Acanthodii (extinct)

Class Actinopterygii
Class Sarcopterygii

Hagfishes

Class Myxini

One order Myxiniformes, with one family, Myxinidae, and 70 species, 3 in Canada (one in Atlantic and 2 in Pacific). Marine, world-wide, temperate regions. Jawless eel-like scavenger feeders that can produce much slime.

Lampreys

Class Petromyzontida

One order, Petromyzontiformes, with 3 families, and 38 species, 10 in Canada (all in Petromyzontidae). Anadromous and freshwater, cool zones of the world. Jawless eel-like fishes, many of which are parasitic-like.

Superclass Gnathostomata - Jawed Fishes

All other fishes and tetrapods are jawed and belong to the superclass Gnathostomata. Two classes are extinct groups of fishes.

Class Placodermi

An extinct group of fishes with bony plates covering the head, living from the Silurian to the Upper Devonian (443 to 360 million years ago); some are known from Canada.

Class Acanthodii

An extinct group of fishes with large eyes and up to 6 pairs of spines between the pectoral and pelvic fins, earliest known jawed fishes living from the Late Ordovician to the Early Permian (458 to 300 million years ago); well known from Canada. Acanthodians are thought to be most closely related to the actinopterygian-sarcopterygian lineage and in classification are usually placed after the class Chondrichthyes.

Jawed Fishes - Class Chondrichthyes/Cartilaginous Fishes

The living members are in 2 subclasses.

Subclass Holocephali

One order, Chimaeriformes, with 3 families (those in North America in family Chimaeridae), and 33 species (a few species in Canada). All marine.

Subclass Elasmobranchii

There are many species of this subclass in Canada, consisting of SHARKS (total of 34 families, 106 genera and 403 species) in 9 orders, Heterodontiformes,

Orectolobiformes, Lamniformes, Carcharhiniformes, Hexanchiformes, Echinorhiniformes, Squaliformes, Squatiniformes and Pristiophoriformes, and RAYS (including skates) (total of 17 families, 72 genera, and at least 534 species), in 4 orders, Torpediniformes, Pristiformes, Rajiformes and Myliobatiformes. Almost all marine.

Jawed Fishes - Class Actinopterygii/Ray-finned Fishes

Actinopterygians, the ray-finned fishes, are placed in 3 subclasses, with, of living members, a total of 44 orders, 453 families, and about 26 890 species.

Subclass Cladistia

One order, Polypteriformes, with one family, the bichirs, of 16 species in freshwater Africa.

Subclass Chondrostei

One order of living members, Acipenseriformes, with 2 families, and 25 species of STURGEONS (freshwater and anadromous), 5 species in Canada, and 2 of paddlefishes (freshwater).

Subclass Neopterygii

Two orders of living members.

- Lepisosteiformes (GARS) with one family of 7 species, 2 in Canada. All usually freshwater.
- Amiiformes (bowfins) with one family of one species, also in Canada. Freshwater.

Jawed Fishes - Division Teleostei

All remaining actinopterygians are placed in the Teleostei, with 4 subdivisions. The Acanthomorpha or spiny-rayed fishes, a branch of the fourth subdivision, contains most of the species of teleosts.

Subdivision Osteoglossomorpha

Two orders of freshwater fishes.

- Hiodontiformes with one family and 2 species (GOLDEYE and mooneye), both in Canada.
- Osteoglossiformes (bonetongues) with 4 families and about 218 species, none in Canada or native in North America.

Subdivision Elopomorpha

Four orders of mostly marine fishes.

- Elopiformes, with 2 families and about 8 species.
- Albuliformes, with 3 families and about 30 species, a few deepwater notacanthids in Canada in Atlantic and Pacific.

- Anguilliformes, with 15 families and about 791 species of EELS, many in Canada.
- Saccopharyngiformes, with 4 families and about 28 species.

Subdivision Ostarioclupeomorpha

Also known as Otocephala, with 2 superorders.

Superorder Clupeomorpha - One order, Clupeiformes, with 5 families and 364 species, eg, in Canada several species in Atlantic and Pacific of anchovies, alewife, HERRINGS, menhaden, SARDINES and shads (also in fresh water in Canada a herring, alewife, shad, and gizzard shad), and elsewhere also pilchards and sprats. Most are marine.

Superorder Ostariophysi - Five orders of mostly freshwater fishes, all except first with modification of first 4 or 5 vertebrae into an apparatus for sound transmission, the Weberian apparatus.

- Gonorynchiformes, 4 families, eg, milkfishes, and 37 species, none in Canada.
- Cypriniformes, 6 families, eg, MINNOWS (50 in Canada plus 4 introductions), SUCKERS (18 in Canada), and loaches, about 3268 species. All freshwater.
- Characiformes, 18 families and at least 1674 species, all freshwater, none in Canada. Popular aquarium fishes.
- Siluriformes, with about 36 families and 2867 species of CATFISHES, normally up to 4 pairs of barbels on head, freshwater on all continents (8 in Canada plus one probable introduction) except Antarctica, and almost 120 species in oceans.
- Gymnotiformes, with 5 families and about 134 species of American knifefishes, eel-like body and electric organs present, freshwater in southern Mexico, Central America, and South America.

Subdivision Euteleostei

Nine superorders, 28 orders, 346 families, and 17 419 species, both freshwater and marine.

Superorder Protacanthopterygii - Four orders.

- Argentiniformes, 6 families and about 202 species of marine smelts, few in Canada over continental shelf.
- Osmeriformes, 3 families and 88 species of freshwater SMELTS (including Southern Hemisphere galaxiids), some in Canada.
- Salmoniformes, one family and 66 species, in 3 subfamilies, GRAYLING (one in Canada), WHITEFISHES and ciscoes (15 in Canada), and CHARS, TROUTS and SALMONS (13 plus one introduction in Canada). Freshwater and anadromous.
- Esociformes, 2 families and 10 species, 4 PIKES, and one mudminnow in Canada. All freshwater.

Superorder Stenopterygii - One order Stomiiformes (dragonfishes), 5 families, and about 391 species, few in Canada mostly in deep water in Atlantic and Pacific. All marine, luminescent organs (photophores) present.

Superorder Ateleopodomorpha - One order Ateleopodiformes, one family, and 12 species, none in Canada. All marine.

Superorder Cyclosquamata - One order Aulopiformes, 15 families, and about 236 species, few in Canada over continental shelf, many in deep water. All marine.

Superorder Scopelomorpha - One order Myctophiformes, 2 families, and about 246 species, many species of lanternfishes occur primarily in deep water off Canada in Arctic, Atlantic, and Pacific. All marine.

Jawed Fishes - Spiny-rayed Fishes

Within subdivision Euteleostei, Division Teleostei, the acanthomorphs or the spiny-rayed fishes. All remaining actinopterygians are placed in this taxon. Many members have true fin spines in the dorsal, anal, and pelvic fins.

Superorder Lampriomorpha - One order Lampriformes, 7 families with about 21 species, in Canada the opah in Atlantic and Pacific and the king-of-the-salmon (a ribbonfish) in the Pacific. All marine.

Superorder Polymixiomorpha - One order Polymixiiformes, one family with 10 species of beardfishes, in Canada one in Atlantic. All marine.

Superorder Paracanthopterygii - An artificial assemblage of 5 orders.

- Percopsiformes, 3 families and 9 species, one in Canada, trout-perch. All freshwater.
- Gadiformes, 9 families and about 555 species, eg, in Canada the marine grenadiers, hakes, CODS, pollock, and walleye pollock, one species confined to fresh water, the circumpolar burbot, including Canada.
- Ophidiiformes, 5 families and about 385 species, eg, in Canada cusk-eels in Atlantic and a viviparous brotula in Pacific and elsewhere pearlfishes.
- Batrachoidiformes, one family and 78 species, toadfishes, one in Canada, the plainfin midshipman in Pacific. Mostly marine.
- Lophiiformes, 18 families and 313 species, anglerfishes, many species with angling device for attracting prey to mouth derived from first ray of spinous dorsal, in Canada in Atlantic a goosefish (called monkfish as a food fish), batfish, frogfish and a gaper, and in deep water in both Atlantic and Pacific several species of ceratioid anglerfishes where miniature males are usually parasitic on the females. All marine.

Superorder Acanthopterygii - The last superorder of actinopterygians contains 13 orders.

- Mugiliformes, one family and about 72 species of mullets, in Canada 2 rarely in Atlantic. Mostly marine.

- Atheriniformes, 6 families and about 312 species of, eg, silversides, in Canada one in Atlantic, one rarely in Pacific, and one in fresh water.
- Beloniformes, 5 families and 227 species of, eg, halfbeaks (in Canada one rarely in Atlantic), flyingfishes (3 rarely in Atlantic and Pacific in Canada), needlefishes (none in Canada), and sauries (in Canada one in Atlantic and one in Pacific). Marine and freshwater.
- Cyprinodontiformes, 10 families and about 1013 species, popular aquarium fishes, eg, killifishes, livebearers, and topminnows, in Canada 3 killifishes in fresh water (one being also brackish in Atlantic) and 2 introduced species into Cave and Basin Hotsprings, Banff National Park.
- Stephanoberyciformes, 9 families and 75 species, in Canada one whalefish rarely in Atlantic and 2 melamphids in Pacific. All marine.
- Beryciformes, 7 families and 144 species, eg, pinecone fishes, roughies, and squirrelfishes, in Canada several species rarely reported in Atlantic. All marine.
- Zeiformes, 6 families and 32 species, eg, dories, in Canada several known in Atlantic and one poorly known in Pacific. All marine.
- Gasterosteiformes, 11 families and 278 species, eg, in Canada pipefishes (one with 3 rare in Atlantic and one in Pacific), seahorses (one rare in Atlantic), tubesnouts (one in Pacific), and STICKLEBACKS (5 as a conservative number, freshwater and marine), and elsewhere also includes cornetfishes and trumpetfishes.
- Synbranchiformes, 3 families, 15 genera and about 99 species, eg, swamp eels, none in Canada. Most freshwater.
- Scorpaeniformes, 26 families and about 1477 species, in Canada many species of SCORPIONFISHES(rockfishes), greenlings, lumpsuckers, poachers, SCULPINS (Atlantic and Pacific and about 10 in fresh water), searobins (rare in Atlantic), and snailfishes. Most marine, worldwide.
- Perciformes, a very diverse group of 160 families and about 10 033 species, in Canada many freshwater species of temperate BASSES (moronids, also in Atlantic and introduced in Pacific), sunfishes (includes largemouth bass) and PERCHES (includes darters and WALLEYE), and many primarily marine species of (larvae and adults that stray into Canadian waters) surfperches, gobies, MACKERELS, TUNAS, swordfish, billfishes, and butterfishes, and elsewhere also freshwater cichlids and primarily marine butterflyfishes, angelfishes and surgeonfishes.
- Pleuronectiformes, 14 families and 678 species, the FLATFISHES

many in Canada, including some of following: flounders, halibuts, plaice, soles, and turbots. Most marine.

- Tetraodontiformes, 9 families and 357 species, in Canada several species occur rarely in Atlantic, and a mola, the ocean sunfish, occurs in Atlantic and Pacific; worldwide, groups include triggerfishes, filefishes, boxfishes, puffers, and porcupinefishes. Most marine.

3

Fish Capture Technology

Fish capture technology encompasses the process of catching any aquatic animal, using any kind of fishing methods, normally operated from a vessel. Use of fishing methods varies, depending on the types of fisheries, and can range from a simple and small hook attached to a line to large and sophisticated midwater trawls or purse seines operated by large fishing vessels. The targets of capture fisheries can include aquatic organisms from small invertebrates to large tunas and whales, which might be found anywhere from the ocean surface to 2 000 meters deep.

The large diversity of targets in capture fisheries and their wide distribution requires a variety of fishing gear and methods for efficient harvest. These technologies have developed around the world according to local traditions and not least technological advances in various diciplines.

In recent decades major advances in fibre technology, along with the introduction of other modern materials, have made possible, for example, changes in the design and size of fishing nets.

The mechanization of gear handling has vastly expanded the scale on which fishing operations can take place. Improved vessel and gear designs, using computer-aided design methods, have increased the general economics of fishing operations.

The development of electronic instruments and of fish detection equipment has led to the more rapid location of fish and the lowering of the unit costs of harvesting. Developments in refrigeration, ice-making and fish processing equipment have contributed to the design of vessels capable of remaining at sea for extended periods.

While these technologies are available, those actually introduced in many small-scale fisheries may amount to no more than motorizing a dugout canoe, use of modern and lighter gear or introducing the use of iceboxes to ensure the quality of the product landed.

The impact of such changes, however, has considerably increased landings and the earnings of fishers, and underlines the need for effective management to prevent excessive fishing effort. The emphasis of much recent technical innovation has been focused on greater selectivity of fishing gear and on gear with less impact on the environment.

ECONOMICS OF FISHING OPERATIONS

Revenues and costs mainly determine the economics of fishing operations. Revenues depend on species and quantities caught and prices obtained, which again depend on marketing channels and markets, seasonal fluctuations and other factors. The main cost factors are capital investment and operation costs, which can be divided in labour costs, running costs and vessel costs. The major components of labour costs are wages and other labour charges such as insurance and employer' s contributions to pensions funds. Running costs are principally composed of fuel, lubricants, cost of selling fish, harbour dues, cost of ice, food and supplies for the crew. The major elements of vessel costs are vessel and gear repair and maintenance expenses and vessel insurance.

In addition, fishing operations also carry external costs which are sometimes difficult to quantify. External costs are defined as costs which are created by a fishing enterprise for others, *i.e.* other enterprises or society, for example through depletion of fish stocks or destruction of the coastal ecosystem. The economic and financial performance of fishing operations is generally assessed with the help of two indicators. For the assessment of the economic performance of a fishing vessel, as in the case of other economic enterprises, the ratio between net cash flow and total earnings (NCF/TE) is used. This ratio is a general indicator of economic profitability/viability of enterprises as it shows the amount of total earnings required by a certain type of fishing vessel in order to generate a given amount of net profit. The financial performance is assessed with the help of the rate of return on investment (ROI). The ratio shows how much money needs to be invested in a fishing enterprise in order to generate a certain net profit.

The economic development and performance of fishing operations is affected by various factors including fluctuations in revenue, perishability of product, falling yields *i.e.* catch per unit of effort, static or falling demand, unforeseen increases in the cost of key inputs and catch and effort restrictions. Theory suggests that in an open-access, unregulated fishery, the fishery will eventually end up producing at the point where total revenue equals total costs. 1999 and 2000, FAO conducted 15 country level studies on the economic and financial performance of marine capture fisheries in close co-operation with fisheries research institutions and national fisheries administrations in selected countries in Asia, Africa, Latin America and Europe.

The findings suggest that, despite fully and sometimes over-exploited fisheries resources, in most cases marine capture fisheries is an economically and financially viable undertaking, which generates sufficient revenue to cover the cost of depreciation as well as the opportunity cost of capital to generate funds for reinvestment in addition to generating employment, income and foreign exchange earnings. Of the 108 types of fishing vessels studied in 15 South American/Caribbean, European, African and Asian countries, 105 or 97 % had a positive gross cash flow and fully recovered their cost of operation. Only three types of vessels *i.e.* stow-nettersinChina and and semi- industrial and industrial shrimp and bottom fish trawlers in Trinidad and Tobagoshowed operational losses. When also considering the cost of capital *i.e.* the cost of depreciation and interest, 92 out of the 108 types of vessels or 85 % showed a net profit after deducting the cost of depreciation and interest.

In the 10 countries, which participated in the previous as well as in the recent study, two countries showed marked improvements in the profitability of their fishing vessels *i.e.* France and Spain while two countries showed a declining profitability *i.e.* the Peoples Republic of China and Germany. In the remaining six countries *i.e.* the Republic of Korea, Indonesia, India, Senegal, Argentina and Peru, the situation was similar as during the previous study carried out between 1995 and 1997.

These overall positive results were also achieved because of higher prices paid to producers as compared to the previous study period. There were only few indications that fishing effort had been reduced and fish stocks had recovered. It was also observed that some fishing fleets had adapted themselves to new conditions dictated by depleted and changing abundance of resources and new access to markets in the context of globalisation by changing their fishing operations.

Those vessels, which had previously shown positive results but now incurred losses, were generally older vessels due to the fact that they continued to work on overexploited stocks. Regarding the impact of government financial transfers it was found that in two countries, belonging to the European Union and in India, almost all types of vessels covered by the cost and earnings study, which received subsidies, would also have been profitable without subsidies. In the Republic of Korea, the situation was mixed while in Thailand, those types of vessels. which could avail themselves of tax exemptions on fuel needed the tax exemption in order to have a positive gross cash flow. No subsidies were available in Indonesia.

In the case of other countries it was observed that no detailed empirical information was available neither on the amount of government financial transfers to the fishing industry nor on the financial performance of individual fishing enterprises. There is a need to collect and analyse information on the quantity and impact of government financial transfers on the economic performance of fishery enterprises with a view to reduce and rationalise the use of scarce government financial resources and to avoid that subsidies have a negative impact on fisheries resources, on the coastal environment and on trade.

Electronic Instruments

The command console on a larger fishing vessel is at the center of the bridge, with a clear view all around. The compass or a repeater is well- displayed for the benefit of the helmsman. The autopilot is built into the steering console with all the associated controls (*i.e.* rudder position, rate of turn indicator). On a fishing vessel the equipment is normally clustered around the skipper's chair but in the largest vessels an additional chair is provided for a co-pilot. In fact, the layout is remarkably similar to the cockpit design of a modern aircraft, with the instruments surrounding the pilot and co-pilot. The main displays are shown on monitors in front of the control position and increasingly these are shown on one large integrated display. Instruments or monitors that are used more infrequently are mounted in the deckhead (*i.e.* ceiling). Probably the only differences from the layout of an aircraft cockpit is that there is more space available on the bridge of a fishing vessel and the bridge windows are tilted downwards rather than upwards.

The instruments spread around the skipper can be grouped into several relatively-loose categories. The first is navigational instruments which are used for the navigation of the vessel while at sea and in harbor. Other instruments are used for fish detection and during the fishing operation, including sonar equipment and electronic aids for the operation. Radio communications are vitally important for the safety and for general communications, and as such are situated around the control position. Communications within the vessel are also provided at this point with links to the messroom, the skipper's cabin, engine room and key points on the working deck of the vessel.

Tension meters built into the winches allow the skipper to monitor the fishing gear and display information, which indicates any malfunction. Trawlershave the tension on both towing warps displayed so that if the gear comes fast on the bottom an immediate alarm is sounded. Differing tensions on the two warps indicate a malfunction unless this is caused by a change in course. In fact, modern computerized winch controls allow for changes in the warp length, to ensure that the trawl remains in a fishing mode throughout a change of course. In purse seines, sensors give the rate of sinking of different parts of the net during setting which can be important. Other types of fishing gear such as gill netsand longlinescan be installed with automatic readouts fitted on the reels and drums which indicate the length of fishing gear which has been set.

CAPTURE FISHERIES

UNCED and its outcomes are said to be a new beginning for international lawmaking, marking the transition from international environment law and international economic law to an international law of sustainable development. In respect of fisheries law, particularly marine capture fisheries, calls on states "to pursue the protection and sustainable development of the marine and coastal environment and its resources" in accordance with the 1982 UN Convention of the Law of the Sea (1982 UN Convention).

After UNCED, the international community adopted a two-tiered approach to deal with the problems of over-fishing. The first was the negotiation of international agreements on specific marine fish stocks and high seas fishing, while the second, which had commenced prior to UNCED, was the development of soft law instruments for the conservation and management of fisheries. Soft law instruments include Chapter 17 of Agenda 21 itself and, most significant, a Code of Conduct for Responsible Fisheries (Code of Conduct), discussed below.

An intergovernmental conference under the auspices of the United Nations to promote effective implementation of the 1982 UN Convention and to take effective action to deter reflagging of fishing boats. Pursuant to this commitment, the UN Conference on Straddling Fish Stocks and Highly Migratory Fish Stocks in 1995 adopted the Agreement for the Implementation of the Provisions of the United Nations Convention of the Law of the Sea of 10 December 1982 Relating to the Conservation and Management of Straddling Fish Stocks and Highly Migratory Fish Stocks (UN Fish Stocks Agreement). At the same time, FAO was working on an agreement to reduce fishing on the high seas contrary to internationally agreed conservation and management measures. In November 1993, the FAO Conference adopted the Agreement to Promote Compliance with International Conservation and Management Measures by Fishing Vessels on the High Seas (Compliance Agreement).

In the meantime, a work plan had been developed in 1992 for the elaboration of a Code of Conduct for Responsible Fisheries. In the run-up to UNCED, a Conference on Responsible Fishing, held in Cancún, Mexico in May 1992, adopted the Cancún Declaration on Responsible Fishing, which called upon FAO to begin development of such a Code. The final text of the Code was adopted in November 1995. The Compliance Agreement was intended to be an integral part of the Code.

The Code of Conduct is intended to cover much more than marine fisheries and high seas fishing (Edeson, 1996). FAO has produced non-legal technical guidelines to provide general advice in support of the implementation of the Code. These include guidelines on fishing operations; vessel monitoring systems; the precautionary approach to capture fisheries and species introduction; the integration of fisheries into coastal area management; fisheries management; conservation and management of sharks; aquaculture development; good aquaculture feed manufacturing practice; inland fisheries; responsible fish utilization; and indicators for sustainable development of marine capture fisheries.

The Code of Conduct was followed by four international plans of action developed under the auspices of FAO: the International Plan of Action for

Reducing Incidental Catch of Seabirds in Longline Fisheries, the International Plan of Action for the Conservation and Management of Sharks, the International Plan of Action for the Management of Fishing Capacity and the International Plan of Action to Prevent, Deter and Eliminate Illegal, Unreported and Unregulated Fishing. Other important international fisheries instruments are the Rome Consensus on World Fisheries, the Rome Declaration on the Implementation of the Code of Conduct for Responsible Fisheries and the Kyoto Declaration and Plan of Action.

The influence of these international law developments and instruments on national legislation cannot be underestimated. Not only have they given impetus at the national level for practical implementation of environmental concepts such as sustainable development and the precautionary principle or precautionary approach in the field of fisheries, but states have also acted to give effect to specific provisions of international fisheries instruments, particularly the Compliance Agreement and the UN Fish Stocks Agreement. Both Agreements seek to enhance responsible fishing or conservation and sustainable use of living marine resources although they emphasize different means to that end.

The Compliance Agreement focuses on ensuring compliance with international conservation and management measures through effective use of flag state responsibility (*i.e.*, ensuring that vessels do not fish on the high seas without authorization). The UN Fish Stocks Agreement reinforces flag state responsibility but emphasizes that conservation and management of straddling and highly migratory fish stocks shall be undertaken through cooperation facilitated largely by regional fisheries bodies (RFBs). This latter requirement, along with the further restriction that non-members of a fisheries management body shall not have access to the fishery resources to which the measures of that fishery management body apply, has had an extraordinary impact in fisheries conservation and management in restricting freedom of fishing on the high seas.

The hard law and soft law instruments adopted in recent years influence and reinforce one another. The preambles of the two Agreements refer to Agenda 21 and to the issues identified in the Cancún Declaration. They also use the term "responsible fishing", which is clearly drawn from the Code of Conduct. Equally, soft law instruments such as the Code of Conduct call on states to become parties to fisheries agreements and to implement them. These connections give a basis for using soft law instruments as aids to interpretation.

AQUACULTURE

Aquaculture is one of the fastest-growing food production systems in the world, and developing countries are the largest producers. Aquaculture offers a number of opportunities to contribute to poverty alleviation, employment, community development, reduction of over-exploitation of natural aquatic living resources and food security in tropical and sub-tropical regions. National legislative action on aquaculture, like capture fisheries, has been similarly influenced by international instruments and developments.

Soft law instruments important to aquaculture include the Kyoto Declaration on Aquaculture, the Bangkok Declaration and Strategy ("Aquaculture Development beyond 2000", adopted at the Bangkok Conference on Aquaculture in the Third Millennium), Chapters 15 and 17 of Agenda 21 and the Code of Conduct.

The first major international Conference on Aquaculture was held in Kyoto, Japan, in 1976. The resulting Kyoto Declaration on Aquaculture was the most important precursor to UNCED in aquaculture, focusing on technology and science, networking, training and institutional development. The Conference was a landmark event that raised awareness of the opportunities offered by, and constraints to, the development of aquaculture.

The Bangkok Conference on Aquaculture in the Third Millennium aimed to reassess the direction of aquaculture development and to propose strategies for sustainable development, with emphasis on regional and global cooperation. "Aquaculture Development Beyond 2000: the Bangkok Declaration and Strategy" is the major output of the Conference. The Declaration states that aquaculture should continue to develop towards its full potential and that policies and regulations should promote practical and economically viable farming and management practices that are environmentally responsible and socially acceptable. National aquaculture processes should be transparent and should take place within relevant national policies, regional and international agreements, treaties and conventions.

Another important soft law instrument draws attention to the special importance of aquatic biological and genetic resources for food and agriculture. Chapter points to aquaculture development as one of the ways forward for coastal states to "obtain the full social and economic benefits from sustainable utilization of marine living resources" in areas under national jurisdiction.

The Code of Conduct sets out principles and international standards with a view to ensuring the effective conservation, management and development of living aquatic resources, with due respect for ecosystems and biodiversity. Article 9.1.1 provides that "States should establish, maintain and develop an appropriate legal and administrative framework which facilitates the development of responsible aquaculture". Though mainly addressed to states, the Code calls upon all those involved in aquaculture to apply it. Governments and the private sector, including interest groups, aquaculture producers and farmers associations, should ensure an enabling environment for sustainable development of aquaculture.

In addition to the principles and guidelines contained in the afore-mentioned instruments, there are regional and technical guidelines that are being developed with a view to ensuring sustainable aquaculture. These include regional guidelines for aquaculture development prepared by the Southeast Asian Fisheries Development Centre; the "Asia Regional Technical Guidelines on Health Management for Responsible Movement of Live Aquatic Animals", developed under the Asia Pacific Regional Aquatic Animal Health Programme and the Beijing Consensus and Implementation Strategy; a supporting manual of procedures; and the "Asia Diagnostic Guide". In 1998, FAO organized the Bangkok Technical Consultation on Policies for Sustainable Shrimp Culture which was followed in 2002 by the FAO/Government of Australia Expert Consultation on Good Management Practices and Institutional and Legal Arrangements for Sustainable Shrimp Culture. The latter produced a set of legal and institutional arrangements for sustainable shrimp aquaculture, which were recommended for adoption by the FAO member states. These arrangements point to the need for a regulatory framework to support the development of sound aquaculture practices.

It is in the area of genetically modified organisms (GMOs) that one finds legally binding instruments of relevance to aquaculture. Globally, more than a dozen transgenic fish are being developed for aquaculture in developed countries. It seems likely that aquatic GMOs will soon be available for sale to consumers. The most significant international action regarding GMOs is the adoption of the Cartagena Protocol on Biosafety under the Convention on Biological Diversity. Its objective is to protect biological diversity from the potential risks posed by the transfer, handling and use of living modified organisms (LMOs) resulting from modern biotechnology. It contains reference to the precautionary approach and establishes an advance informed agreement procedure for ensuring that countries are provided with the information necessary to make informed decisions before agreeing to the import of such organisms into their territory.

Finally, as discussed, the Agreement on Sanitary and Phytosanitary Measures (SPS Agreement) of the WTO recognizes the *Office international des épizooties* (OIE) as the relevant international organization responsible for the development and promotion of international animal health standards, guidelines and recommendations affecting trade in live animals and animal products. At present the OIE International Aquatic Animal Health Code is the only source of international standards recognized under the SPS Agreement for health certification requirements for international trade in fish and shellfish. Although OIE standards are not binding in themselves, countries that become members of the WTO are bound to follow international standards where they exist.

LEGISLATIVE TRENDS IN CAPTURE FISHERIES

Over-fishing has led to the collapse of many of the worlds fisheries, in many instances because of the existence of open access regimes where the fisher chooses where and when to fish and how much fish to take. Such regimes were justified largely on the erroneous belief that the oceans and other water bodies hold infinite fish resources. In addition, the oceans, in particular the high seas, were *res communis* so that everyone had a right to fish and no one had ownership over the resource or the right to limit access. While many large areas of oceans are now subject to the jurisdiction of coastal states, and high seas fishing is subject to rights and obligations under the 1982 UN Convention, one may be surprised to find that many states have only recently begun to regulate domestic fishing or to consider stringent control over fishing.

In many Southeast Asian states located to the West of the South China Sea, for example, an individual has an unquestionable right to take to the sea to fish, provided that he or she complies with whatever requirements or regulation is in place. In Thailand, an act such as registration of a fishing vessel which is a prerequisite for entering the coastal marine fishery is a mere formality. In this situation, the fisheries management authority in general considers its role to be one in the service or support of fishers or for the promotion and development of fishing.

In Tonga, fishing, particularly small-scale fishing, is considered a livelihood, and no licences have been issued for medium- to large-scale commercial fishing until recently. In many cases government regulation is seen as an unnecessary hindrance to the right to fish. However, the trend today is that governments are realizing that in order to ensure the long-term sustainability of fishery resources, the once open access regime for fishing can no longer continue. To this end, Thailand and Tonga, among others, are considering legislation which could make the right to fish subject to significant controls. Given that it could still be a matter of opinion whether anyone has a right to fish in certain jurisdictions, governments like Iceland enacted fisheries legislation that attempts to clarify the issue as follows:

"Marine resources that are found in Iceland waters and are utilized are the common property of the Icelandic nation. The purpose of this legislation is to ensure the preservation of and sensible utilization of these resources thereby guaranteeing full employment and stable settlement of the country. The issuing of fishing permits, in accordance with this legislation, does not constitute any claims to ownership or irrevocable claims by individual parties over fishing rights".

Senegal makes a similar statement in its 1998 fisheries legislation. Even countries that have never before regulated fisheries, such as Ethiopia, are considering new policies and legislation to properly manage the significant increase in fishery activities and to prevent over-exploitation. As a land-locked country, Ethiopia's fishery activities occur entirely in inland water bodies such as rivers, lakes and reservoirs. The proposed fisheries laws for Ethiopia set out the basic framework for fisheries management and contain many of the principles referred to above, including the precautionary approach.

Other jurisdictions have focused more on asserting the right of the government to manage fisheries resources. Thus Namibia recently stated in legislation that the management, protection and utilization of marine resources in Namibia and Namibian waters shall be subject to its Marine Resources Act of 2001. Cameroon and the Marshall Islands set out the right of the government to manage and control the fisheries resources in stronger terms. A similar approach, in the context of inland fisheries, can be found in Malawi's legislation.

TOWARDS THE USE OF PROPERTY RIGHTS

Related to the trend away from open access to limited access regimes is the move towards creating property rights in fisheries resources and the allocation of such rights. The move from an open access to a limited access regime is in essence a move from one form of administration of property rights to another.

A property rights regime can be a state property regime, private property regime or common (collective) property regime. The most universal is the state property regime, although most states do not commonly depend on "property" or know and administer it as such, but rather on the distribution of fishing rights on the basis of jurisdiction or sovereignty. In such a regime, the state is the custodian of the fishery resource and can decide to leave the resource to free use (open access), can choose to exploit the resource directly through its own agencies or can allot rights to citizens to exploit the resource. The most widely used and practical system of administering the last variant is that the state grants licences to individuals or groups to fish, and in this way controls access to the resources. At the other end of the scale is the common property regime whereby a local community instead of an individual holds exclusive rights to harvest fish in a certain geographical area.

The developments in Eastern Europe and the former Soviet Union illustrate the effects of the different property regimes. Before the shifts to market economies, fisheries were maintained at an almost constant level by centralized economic plans. However, during the transition period, the fisheries operated under an open access system which stimulated competition but at the same time increased the risk of over-exploitation of the stocks. Therefore, new fishing regulations have been established that set out fish licensing and catch quota regimes, to minimize the risk of collapse. For example, Lithuania enacted a Law on Fisheries in 2000.

Under the private property regime, normally the authorized user who has received a licence has a personal right to fish or harvest and such right is renewed regularly. More recently, licensing schemes have begun to permit the authorized user to sell or lease the right to fish. Such a property regime requires the setting of the total allowable catch (TAC) for a specific fishery, and establishes systems for allocation of the TAC, the transferability and leasability of the rights and the manner in which such rights can be enforced. Iceland has institutionalized the individual quota system in legislation, as has New Zealand. In Africa, Angola and Mozambique have legislative provisions that will enable the use of a property rights system in the future.

FROM DEVELOPMENT TO SUSTAINABLE UTILIZATION

Fisheries management prior to Rio generally emphasized optimum utilization of fishery resources. In respect of marine capture fisheries, this perspective is reflected in the 1982 UN Convention, which calls on states to optimally utilize fisheries resources. Many fisheries management authorities at that time focused on building capacity or encouraging entry into the fisheries industry to increase fish production for individual profit, revenue generation or domestic consumption. Fisheries management was also an exclusive mandate, often devoid of considerations of the impact of fishing or fisheries activities on associated matters such as the environment. Conversely, other industries did not consider the impact of their activities on fisheries.

There has been a gradual shift in focus to a more "rounded" approach to fisheries management - one that not only ensures exploitation of the resource for economic gain, but also ensures that the resource is maintained at biologically, environmentally and economically sustainable levels. The shift in focus which has been forged in international fora is slowly becoming entrenched in domestic policy and eventually has found its way into legislation. A few examples demonstrate this. Whereas in the past the statement of purpose in fisheries legislation might refer to allocation of a quota levy, the development of fishery or the management of fishery, in recent times one sees an inclusion of a statement of principles and policy relating to sustainable utilization of fishery resources, as in Namibia, Nauru, New Zealand, Papua New Guinea and South Africa. For example, the Marine Resources Act of Namibia has as its purpose to "provide for the conservation of the marine ecosystem and the responsible utilization, conservation, protection and promotion of marine resources on a sustainable basis; for that purpose to provide for the exercise of control over marine resources; and to provide for matters connected therewith". The South Africa Marine Living Resources Act of 1998 has a similar provision.

While preambular provisions of a fisheries act are usually not enforceable, these statements of purpose are given effect in the law's operational provisions, *i.e.*, in the mechanisms it establishes. More significantly, the strong reference in such laws to conservation of the ecosystem and to long-term sustainable utilization is important for it stamps them as very much of the new era of fisheries management.

Similarly, management action was, in the past, based primarily on scientific (biological) information directly affecting the species or fish stock in question. Little or no consideration was given to the effect of fishing on species associated with or dependent on the target species or on the aquatic environment. National legislation now calls for the effects of fishing on non-target or associated species or the aquatic environment to be considered in determining the types of management measures that should apply in a fishery, and to what extent. Legislation that demonstrates this trend can be found in New Zealand and South Africa. Laws of other jurisdictions may not necessarily refer to environmental protection, as in New Zealand, or allow for environmental impact assessments, as in South Africa. Nevertheless, other management actions required under fisheries laws may have the effect of protecting the environment. Sustainable use of the fish stocks and the protection of the environment are now key elements in the recently developed legislation and institutions of Albania, Hungary, Lithuania and Romania.

A related and interesting feature is an increase in references to the precautionary approach or principle. Such references are viewed as an effort by states to give effect to international fisheries instruments or agreements, particularly principle 15 of the Rio Declaration, the UN Fish Stocks Agreement and the Code of Conduct. Even if the precautionary approach is only mentioned in a preambular provision or a broad policy statement and may at best be used only as an aid to interpretation, its significance should not be underestimated, for it "represents a major change in the traditional approach of fisheries management, which until recently has tended to react to management problems only after they reached crisis levels".

IMPROVED ENFORCEMENT

As countries seek to ensure that only sustainable levels of fishing activity are allowed in zones under national jurisdiction, several innovative mechanisms and legislative approaches to fisheries monitoring, control and surveillance have emerged, in particular to curb illegal, unreported and unregulated (IUU) fishing.

Vessel Monitoring Systems

The use of satellite-based vessel monitoring systems (VMS) is a recent development in fisheries monitoring control and surveillance (MCS). VMS ensures that fishing vessels provide reports in real time. VMS can be seen as a direct response to IUU fishing, in particular fishing that is unreported due to problems with radio reporting systems and other conventional means of reporting vessel positions.

At this time, VMS is focused on position reporting, although other VMS information, namely, sighting reports, catch reports, notifications (entry/exit into the exclusive economic zone, port entry, etc.) and analyses, can also be generated by VMS. VMS is currently considered a complementary tool to conventional MCS tools such as sea and air reconnaissance. VMS is in use or in various stages of trial and implementation in many countries and by regional fishery bodies. In the case of one regional fishery body, the Forum Fishing Agency (FFA), the member countries that are mostly developing countries sought to overcome their individual limited resources for MCS by establishing a regional VMS.

The issues in the implementation of VMS which are dealt with in legislation are: requiring the installation of VMS components, namely automatic location communicators or vessel tracking units; protecting VMS components; ensuring confidentiality of VMS information; and using VMS information in fisheries enforcement in courts, particularly the admissibility of VMS information such as vessel positions. Recent legislation on VMS in Europe, North America, Southern Africa and the South Pacific has sought to address these issues.

"Long-arm" Approach to Enforcement

Another mechanism that has emerged in fisheries law enforcement is a provision in national fisheries legislation commonly referred to as the "Lacey Clause", from the Lacey Act of the United States. The provision extends the arm of the law, basically by making it unlawful to import fish that has been taken contrary to the laws of another country. (Kuemlangan, 2000) A common example of violation of the laws of another state is the taking of fish without a licence where such licence is required by that states' fisheries legislation.

A typical Lacey Act provision states that anyone who lands, imports, exports, transports, sells, receives, acquires or purchases any fish taken, transported or sold contrary to the law of another state is guilty of an offence and liable to a fine. Such a clause was first adopted in the FFA region by Papua New Guinea in 1994, followed by Nauru in 1997 and Solomon Islands in 1998. New Zealand has introduced in a recent amendment to its principal fisheries legislation a Lacey Act-type clause, which prohibits its nationals from fishing in another jurisdiction in contravention of that jurisdiction's laws. An interesting aspect of the New Zealand legislation is that the prerequisite element of bringing fish into the country is not necessary, although such extra-territoriality applies only to New Zealand nationals and vessels.

Alternatives to Criminal Proceedings

Despite innovative efforts such as enactment of Lacey Act clauses, enforcement of fisheries provisions through criminal laws and procedures has a number of drawbacks. There is a high standard of proof, and there may be difficulties in using evidence generated by VMS due to the hearsay rule. In many jurisdictions, extended delays plague the criminal law system. One solution, now applied in the United States, the FFA region and many civil law countries, is the adoption of civil and administrative processes and penalties for dealing with fisheries offences. This approach presents the advantages of expedited proceedings, lower standards of proof, possibilities for negotiated settlements and hearings which do not necessarily follow strict rules of evidence. Civil penalty schemes for fisheries violations treat certain violations of fisheries laws as civil wrongs penalized by civil penalties, while the right of the offender to decide that he or she be tried under the normal judicial process is preserved.

While only a few countries have adopted administrative proceedings to deal with fisheries offences, many states in the Caribbean, Indian Ocean and South Pacific regions have adopted a system of compounding of offences in order to deal swiftly with fisheries violations. The main element of such a scheme is that instead of resort to traditional court proceedings, the person in whom powers to compound offences is vested (usually the Minister responsible for fisheries or the chief executive officer in the fisheries administration) decides to accept sums of money - usually not more than the maximum of fines allowed - from the offender if it is believed that an offence has been committed. Other requirements in more recent legislation are that offences may be compounded only with the consent of the alleged offender and that the Minister or chief executive officer may be empowered to release any article seized in relation to the offence. The offender retains the right to have the matter against him or her heard in normal judicial fora.

REQUIRING FISHERIES MANAGEMENT PLANNING

Fisheries management planning involves the development of a plan which describes a particular fishery and its particular problems, the actors involved in the fishery and the objectives for the development of the fishery, and which outlines the measures for control of fishing to ensure the sustainable utilization of the fishery resource. The management plan may be linked to mechanisms that limit entry or restrict fishing. For example, some of the provisions might provide that a licence may be refused on the grounds that to issue it would undermine the objectives of a fishery plan. The fisheries management plan should be developed with input of technical experts, and provisions should be included to monitor its implementation.

For many countries, the development of a fishery plan used to be an administrative process and the prerogative of the fisheries management authority. In such countries, fishery management planning was a good management technique but not a mandatory requirement. In recent years, however, more countries have embraced the management planning concept and have eventually required the development of fishery plans in legislation. The Australian state of Queensland prescribed the development and implementation of management plans for the state's major fisheries in legislation in 1994. Papua New Guinea first introduced the use of fishery plans in the same year. Nauru introduced its version of fishery plans, the "fisheries strategy", in 1997. Solomon Islands legislatively introduced fishery management and development plans for the first time in 1998. Malawi and Senegal also legislatively introduced fisheries management plans in legislation in 1997 and 1998, respectively. These are but a few of the many countries that have adopted the use of fishery plans in fisheries management in the last decade.

INCREASED PARTICIPATION AND DEVOLUTION OF FUNCTIONS

National fisheries management, like many natural resource management regimes, heavily concentrated management authority in the central (national) government or its agencies. In recent years, however, the effectiveness of these top-down management approaches has been questioned.

The main criticisms include:

- The lack of consultation with stakeholders or with the regulated, which results in the lack of a legitimate basis for regulations or management measures and, consequently, non-compliance;
- Implementation and enforcement in such systems relies heavily on adequate technical capacity and other resources, of which there are never enough, thus adding to the high rate of non-compliance;
- Resource users do not appreciate that the sustainable use of the fisheries resources is vital to their livelihood or the nation's economy because they do not feel part of or "owners of" the management process and measures.

Recent legislation has expanded the scope of involvement of stakeholders so that there is broader participation both at the decision making and implementation levels. The expansion of involvement of stakeholders has been in three main areas:

- Consultation, whereby the management authorities solicit the views of persons who are interested in or could be affected by the management decision, so that their views can confirm or cause an amendment to the proposed management decision or regulation as appropriate;
- Formal representation of stakeholders on consultative, advisory or decisionmaking institutions within the fisheries management framework;
- Devolution of management or implementation powers, or both, to lower-level governments and stakeholder communities or groups. Examples of the expansion of participation in the consultation process can be found in the fisheries legislation of Barbados, New Zealand and the Philippines. In the context of fisheries management plans, it is argued that the participation of stakeholders in the preparation of such plans increases the opportunity for their effective implementation.

Broad participation through formally established institutions is a more common practice than participation through direct consultation or devolution of powers. Barbados and Mauritius facilitate such participation through Advisory Committees, whereas Malawi has an Advisory Board and South Africa has Consultative Advisory Forums. In formerly communist regimes, Albania established Central and Local Consultative Commissions whereas Lithuania has a Fisheries Board.

Participation of stakeholders by means of the third approach (*i.e.*, devolution of powers) is a more progressive form of participation in fisheries management compared to the other two.

It is also the most difficult to institutionalize because it involves laws relating to political governance which may implicate the constitution or other fundamental laws on governance or decentralization. Nevertheless, recognition of the important role to be played by communities whose livelihoods depend on fishing, and transfer of fishery management functions to such communities or to lower-level governments, has been slowly gaining popularity (as well as controversy) over the last decade. Canada, New Zealand and the United States, which have been negotiating with native or aboriginal communities for years, have developed schemes to recognize or give deference to native rights to fish or manage fisheries. In Japan, fisheries cooperatives have enjoyed some form of exclusivity in fishing and management over coastal resources falling within their domain.

The Philippines has been a leader in the developing world in the devolution of fisheries management powers. The government formalized the decentralization of fisheries management powers to municipalities in 1991 through legislation and consolidated it in subsequent legislation in 1998. In the Marshall Islands, Local Government Councils are responsible for the management, development and sustainable use of the reef and in-shore fisheries, extending up to five miles seaward from the baseline from which the territorial sea is measured.

A variant of devolution regimes is co-management, as seen in Malawi. There, since 1993, participatory fisheries management programmes have been introduced whereby local-level institutions and the Department of Fisheries jointly make decisions. This scheme has replaced the former centralized management system.

FOOD SAFETY REGULATION FOR FISH AND FISH PRODUCTS

In many developing countries, seafood exports are an important source of revenue and there is growing interest in government policies and practice relating to fish and fish product exports. Although food safety is generally the purview of ministries or agencies concerned with food and human health, in some countries government may decide that seafood safety is best regulated by the authorities responsible for the fisheries sector, and this may require legislative action.

Food safety issues affect fisheries particularly in the post-harvest sector. The main trend is the introduction of the Hazard Analysis and Critical Control Point (HACCP) system into the European Union (EU) and the United States, and its recognition as a food safety assurance mechanism by Codex Alimentarius. The EU and the United States have both made fish and fishery products the first category of foods in the food industry subject to mandatory application of HACCP systems.

The EU issued the first regulation for fish products "laying down the health conditions for the production and the placing on the market of fishery products" in 1991 (Directive 91/493). In May 1994, the EU adopted an additional regulation which made it mandatory to impose more precise rules for the application of health checks. The United States adopted a seafood HACCP regulation, Procedures for the Safe and Sanitary Processing and Importing of Fish and Fishery Products, in December 1997.

These developments have caused the major seafood-exporting countries to adopt or meet the standards established by the importing countries. For example, Namibia and South Africa, which have significant exports to the EU, require that seafood processors and exporters meet Directive 91/493. The directive concerns both domestic (EU) and third country (non-EU) production. It defines EU standards for handling, processing, storing and transporting fish. It must be noted that processed bivalve molluscs (as well as tunicates, marine gastropods and echinoderms) are subject to both Directive 91/492 and Directive 91/493.

Directive 91/493 lays down rules on conditions applicable to factory vessels, on-shore plants, packaging, storage and transport. Provisions that may require more details are set concerning auto-controls, parasites (all visible parasites must be removed), organoleptic, chemical and microbiological checks. National authorities responsible for standards, food safety and fisheries management are also charged with encouraging exporters to meet HACCP and ISO 9000 standards and guidelines, with the ultimate objective of formally adopting these systems through regulation.

In Tonga, seafood exporters to the United States have taken it upon themselves to implement HACCP and other applicable requirements of the seafood safety regulations administered by the U.S. Food and Drug Administration.

It is a matter of Tonga government policy that regulations for the processing of seafood and a certification system for seafood exports are to be introduced soon. To this end, draft seafood safety regulations are currently under consideration for promulgation under the principal fisheries legislation. The United Republic of Tanzania is one of the few countries to adopt regulations (the Fish Quality Control and Standards Regulations of 2000) to specifically implement HACCP.

LEGISLATIVE IMPLEMENTATION OF INTERNATIONAL FISHERIES INSTRUMENTS

The 1993 FAO Compliance Agreement and the 1995 UN Fish Stocks Agreement address the nature of state obligations at the international level; both agreements seek to define with a degree of specificity which is unusual in global fisheries agreements the precise ways in which state parties should meet those obligations. Implementing legislation will be necessary in most countries to meet these requirements as a precondition to state ratification or acceptance. This will usually call for new legislation, although, depending on the legislation already in place, some countries may be able to implement the Agreements through changes to current legislation or through subsidiary legislation.

COMPLIANCE AGREEMENT

The Compliance Agreement reinforces the effectiveness of international fisheries conservation and management measures by redefining and reinforcing the concept of flag state responsibility for the activities of fishing vessels flying the flag of a state party. It also seeks to provide means to ensure the free flow of information on all high seas fishing operations.

The agreement requires the following issues to be implemented in the national legislation of its parties:

- Designation of the national authority responsible for carrying out the duties of the flag state under the agreement;
- Provisions that make it unlawful for flag vessels to undermine the effectiveness of international conservation and management measures, and that provide a mechanism for authorities to ensure that the law is respected;
- Mandatory fishing authorizations for flag vessels fishing on the high seas;
- Mandatory conditions for flag vessels receiving a fishing authorization;
- Proper marking of fishing vessels;
- Information on fishing operations;
- Enforcement measures and sanctions;
- Establishment and maintenance of records of flag vessels fishing on the high seas;
- Duties of the flag state to provide FAO with information.

The Compliance Agreement has not yet entered into force but the parties to the agreement have started to give effect to it through legislation. At the time of writing, the following countries had enacted legislation or made amendments to existing legislation to implement the provisions of the agreement: Australia, Canada, Namibia, New Zealand, Norway, Seychelles, South Africa, Saint Vincent and the Grenadines and the United States. The style with which these countries have legislated on the essential components of the Compliance Agreement, and the scope of their legislative action, varies as the countries attempt to reflect their national situations and respond to their particular needs. Time does not permit an examination of the details of these national efforts, although it can be said that these countries have clearly charted the general approach on how to incorporate the requirements of the agreement into national legislation.

PRECAUTIONARY APPROACH TO FISHERY TECHNOLOGY

OBJECTIVE

Recognising that many aquatic resources are overfished and that the fishing capacity presently available jeopardise their conservation and rational use, technological changes aimed solely at further increasing fishing capacity would not generally be seen as desirable.

Instead a precautionary approach to technological changes would aim at:

- Improving the conservation and long-term sustainability of living aquatic resources;
- Preventing irreversible or unacceptable damage to the environment;
- Improving the social and economic benefits derived from fishing, and
- Improving the safety and working conditions of fishery workers.

INTRODUCTION

Fishery technology consists of the equipment and practices used for finding, harvesting, handling, processing and distributing of aquatic resources and their products.

Different fishery technologies will have different effects on the ecosystem, the social structure of fishing communities, the safety of fishery workers and the ease, effectiveness and efficiency of management of the fishery. It is the amount and context in which fishery technology is used (*e.g.*, when, where and by whom) that influence whether the objectives of fisheries management are reached, and not the technology. For instance, the current overfishing of many aquatic resources is the product of both the efficiency of the finding and catching technologies and of the amount used. Similarly, building a fishmeal plant might involuntarily result in severe changes in the way the fishery is conducted, and in the community's social structure.

Fishery technology is constantly evolving and its efficiency in catching fish will increase over time. For example, a 4 per cent increase in efficiency per year would cause a doubling of the fishing mortality rate in 18 years if the fishing effort remained constant. A precautionary approach to management should take such increases into account.

A precautionary approach should be adopted for the development of new technologies or the transfer of existing technologies to other fisheries to avoid unplanned abrupt changes in fishing pressure or social structures. Certain technologies will be considered undesirable, if they create unacceptable effects (*e.g.*, poison or explosives) or if their adoption leads to wasteful use (*e.g.*, at sea, sorting machines have been banned where they might increase discarding).

Fishery technologies produce side effects on the environment and on non-target species. These effects have often been ignored but, in the context of a precautionary approach, some technologies may warrant a review. Similarly, a precautionary approach would encourage careful consideration of the side effects of new fishery technologies before they are introduced.

Each fishery technology has advantages and disadvantages that should be balanced in a precautionary approach, and it may be better to have a mixture of technologies. When new fishery technology is introduced, it should be carefully evaluated to assess its potential direct and indirect effects. If a mix of fishery technology representing "best current practice" in an area can be identified, precautionary management would encourage its adoption while it would discourage damaging ones. Responsible fishery technology achieves the specific fishery management objectives with the minimumdamaging side effects. These concepts (of responsible fishing and best current practices) were addressed by the UN General Assembly and in the Cancun Declaration.

A precautionary approach would provide for a process of initial and on-going review of the effects of fishery technology as it is introduced or evolves in local practice. However, the extent to which a precautionary approach can be applied to the management of technological changes depends on the existing level of management. In some cases, education of fishermen and consumers towards responsible practices may be the only possible approach. Where elaborate research, management and enforcement systems are in place, a wider variety of options are available for application of the precautionary approach. However, although some gears and practices are prohibited they may continue to be used. The adoption of a precautionary approach to the management of new fishery technology depends on the ability to achieve compliance through education and/or enforcement. The following sections assume that institutional arrangements exist to achieve compliance.

EVALUATING THE IMPACTS OF TECHNOLOGIES

A precautionary approach to developing and selecting responsible technologies for fishing requires an appropriate understanding of the consequences of their adoption and use. These consequences, particularly the impacts on non-target species and ecosystems, may be highly uncertain. Nevertheless, some information exists and more can be obtained. The problem of evaluating impacts is relevant both to the use of existing technologies and to the development of new ones, as well as to the introduction of existing technologies to new areas. The description of a given technology would state its relative impacts and advantages for a given species in a specific environment. Target fishery, environmental and ecosystem, socio-economic and legal factors should be considered when evaluating the impacts of fishery technologies.

The factors to consider when evaluating the impacts of fishery technology include:

- Target-fishery factors such as selectivity by size and species (*e.g.*, target, non-target, and protected species; discards; survival of escapees; "Ghost fishing"; and catching capacity);
- Environmental and ecosystem factors such as bio-diversity; habitat degradation; contamination and pollution; generation of debris and rubbish disposal; direct mortality; predator-prey relationships;
- Socio-economic factors such as safety and occupational hazards; training requirements; user conflicts; economic performance; employment; monitoring and enforcement requirements and costs; and techno-economic factors (*i.e.*, infrastructure and service requirements; cost and technological accessibility; product quality; and energy efficiency), and
- Legal factors such as existing legislation; need for new legislation; international agreements; and civil liberties.

These factors could be used to identify beneficial new technologies or damaging ones, to assess the ability of a fishery to accommodate increased use of an established technology and to help direct monitoring and special reporting procedures towards important questions. Technologies for aids to navigation, fish-locating devices, processing and distribution could also be described and evaluated using the above criteria. This will require a suitable description of technologies, cross-referenced against a range of possible impacts. Other elements relevant to the specific technology/area evaluated would also be included.

The approaches used to evaluate impacts will vary according to the human and financial resources available to collect information. If resources are limited, it may be possible to make decisions based on existing information on the impacts of similar technologies in similar environments. Monitoring of existing fishing practices (for example recording of bycatch) will provide additional information.

Where financial and human resources are limited, existing information on impacts could be used to do desk studies following the approach to evaluation suggested above. Although some general guidelines can be given, based on known characteristics of types of resources and technology, the most appropriate mix of technologies to be used in a particular fishery should be established on a case-by-case basis, following evaluations made at appropriate regional and national levels. Such evaluations could be refined with practical experience and weighed in accordance with local social and economic values.

In the case of new technologies, or technologies new to an area, pilot studies may be cost-effective in evaluating the impacts and can be useful in demonstrating the benefits of new technology. For example, the introduction of escape ports in lobster traps for undersised individuals demonstrated to fishermen that catch rates of large lobsters increased. On the other hand, pilot studies cannot demonstrate long-term gains such as increased yield per recruit, but they will show the short-term losses.

Considerable resources are required for major experiments to measure effects of fishery technology on the marine environment, but well-designed experiments of this type (either as research projects or via experimental management) will provide the most useful information on which to judge the impacts of technologies in particular areas or habitats. This information may be relevant in other areas than the study sites or fisheries from which the data were derived.

Procedures developed in other contexts for protecting the environment could also be suitable when evaluating new technologies in fisheries or major alterations to existing ones. This would be particularly necessary when there are vulnerable resources or fragile ecosystems, that must be protected. In a precautionary approach, proponents of new fishery technology would be required by the State to provide for a proper evaluation of the potential impacts of new techniques before authorisation is given.

The maximum cost that could be justified for evaluating new fishery technology or practices should be in proportion to the expected benefits and impacts.

IMPLEMENTATION

In a precautionary approach to managing fishery technology, a designated lead authority should have the mandate to evaluate and decide on the acceptability of a proposed new technology, or changes to existing technology, and oversee the impact evaluation procedure. Proponents and other stakeholders should be able to appeal if the proper procedure has not been followed or if the decision by authorities does not appear to agree with the conclusions of the review. As authorisation procedures in the majority of cases would be for minor technical improvements, the procedures could be kept simple and administration costs held at a relatively low level.

However, minimal progressive improvements will accumulate over time and periodic reviews of the impacts of existing technology will be necessary. Increases in catching efficiency result from the rapid growth in the use of modern information technologies in most fisheries around the world (acoustic fish detection and identification, gear and vessel monitoring, satellite-based environmental sensing and navigation, and easy inter-vessel communication). However, information, formally treated as "a measure of the reduction of uncertainty", can also potentially improve selectivity, safety and profitability of fishing operations and thus create beneficial effects.

Restricting the use of improved information technologies will rarely be justified or successful and there should be a positive attitude towards technical progress in fisheries in general especially with regards to safety at sea and fishermen's health.

The benefits of technological improvements need adequate extension work and education to encourage their adoption. The promotion of the best technology would benefit from improvement in international cooperation regarding technology transfer, as underscored in UNCED's Agenda 21. The successful international efforts in the Eastern Central Pacific in training crews in effectively avoiding bycatches of dolphins through the use of specifically designed technology is a good example of what can be achieved in this respect.

TECHNOLOGY RESEARCH AND DEVELOPMENT

Fishery technology research in support of a precautionary approach would encourage the improvement of existing technologies and promote the development of appropriate new technologies. Such research would not just concentrate on gears used for capture; for example, research into the cost-effective purification of water supplies to ice plants might considerably reduce post-harvest losses and improve product quality and safety.

Technological developments such as satellite tracking may also help precautionary management by improving monitoring of commercial operations and by enabling research to reduce uncertainty about relevant aspects of fisheries science.

IMPLEMENTATION GUIDELINES

The following measures could be applied in order to implement a precautionary approach to fishery technology development and transfer.

Authority

- Effective mechanisms to ensure that the introduction of technology is subject to review and regulation should be established.

Evaluation Procedures

- A first step in the evaluation procedure is the documentation of the characteristics and amount of the fishery technology currently used.
- Procedures for the evaluation of new technologies with a view to identify their characteristics in order to promote the use of beneficial technologies and prevent usage of those leading to difficult-to-reverse changes should be established.

- These procedures should evaluate with appropriate accuracy the possible impacts of the proposed technology in order to avoid wasteful capital and social investments.
- Authorities should ensure that proponents and other stakeholders understand their obligations and their rights regarding such procedures.
- The extent of the evaluation procedures should match the potential effects of the proposed technology, *e.g.*, from desk study through full scale impact studies, possibly including or leading to pilot projects.

Implementation

- Authorities should implement technology gradually to minimize the risk of irreversible damage or overinvestment.
- Existing technologies and their effect on the environment should be reviewed periodically.
- Technological developments may modify the practices of fishery workers. To achieve the full benefits of the technology and to ensure the safety of fishery workers, training in the proper use of the new technology should be provided.
- In fisheries that are being rehabilitated, the opportunity should be taken to review the mix of technologies used.
- Research into responsible fishery technology should be encouraged.
- Technology research for the reduction of uncertainty in stock assessment and monitoring should be encouraged.

TYPES OF OVERFISHING

GROWTH OVERFISHING

Growth overfishing occurs when animals are harvested at an average size that is smaller than the size that would produce the maximum yield per recruit. The total yield from the fishery is therefore less than it would be if the fishing mortality rate, or percent of the stock removed each year, was lower. In such a case, less fishing would produce higher landings. Growth overfishing still reduces the potential yield from a fishery, and thus the economic and other benefits that could be obtained from the stock.

For example: Heavy fishing removes the larger animals and does not give young lobsters a chance to grow to their potential size. Most of the large lobsters in the inshore fishery were caught during the 1800s and early 1900s. Lower fishing pressure on the offshore lobster stocks allows more lobsters to grow to a larger size, but even there the size range is much reduced from the early days of the fishery.

The largest lobster ever recorded was a 51 1/2 pound lobster caught in Marine in 1926.

Recruitment Overfishing

The rate of fishing above which the recruitment to the exploitable stock becomes significantly reduced. This is characterised by a greatly reduced spawning stock, a decreasing proportion of older fish in the catch, and generally very low recruitment year after year. Recruitment refers to the time when young fish become big enough to be caught. Recruitment overfishing means so many adults have been caught that there are no longer enough young adults to maintain the population. When fishing pressure is too heavy to allow a fish population to replace itself. Occurs in circumstances where an increase in fishing effort from current levels will cause a decline in recruitment to a fishery.

Recruitment overfishing can be brought about by:

- Reduction of the spawning stock (which may become so small as to produce a limited number of eggs and hence of recruits), and

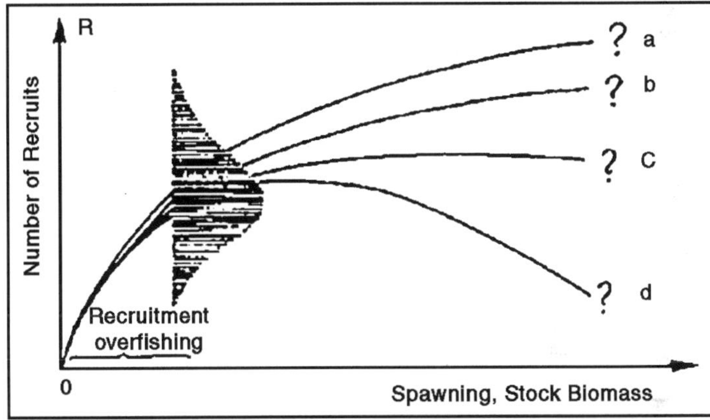

Fig. The Stock-recuitment Relationship in Connection with the Concept of "Recuitment Overfishing"

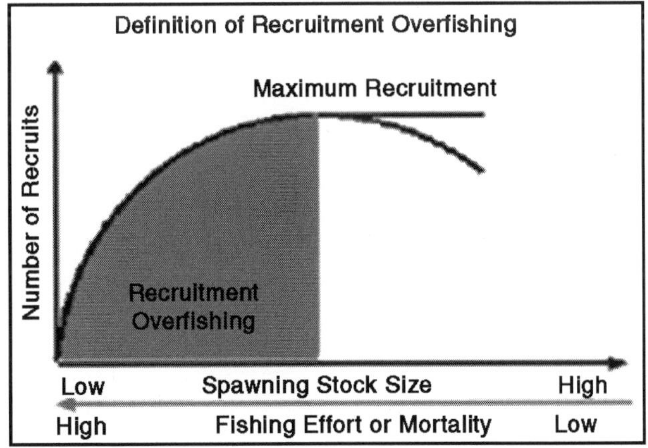

Fig. In the Classical Sense, a Level of Fishing Effort or Fishing Mortality Such that a Reduction of this Level Would, in the mediun term, Lead to an Increase in the total Catch.

- Coastal environmental degradation, which affects recruitment through its effects on the size and/or suitability of nursery areas. [Note that preventing recruitment overfishing is not, as[some] think, a matter of letting 'each female spawn at least once', since, for example, less than one in a thousand anchovy or shrimp larvae reach a mature age, even in the absence of a fishery. Instead the crucial aspect is that the spawning stock should be large enough to ensure that the subsequent recruitment remains independent of the parental stock.]

Ecosystem Overfishing

Occurs when the species composition and dominance of an ecosystem is significantly modified by fishing, *e.g.*, with reductions of large, long-lived, demersal predators and increases of small, short-lived species at lower trophic levels. Competition and predation between taxa. It may be described here as what takes place in a mixed fisheries when the decline (through fishing) of the originally abundant stocks is not fully compensated for by the contemporary or subsequent increase of the biomass of other exploitable animals. Overfishing is the biggest threat to the ocean environment.

Concern for biodiversity stems finding that overfishing of higher-trophic-level fish stocks (*i.e.*, piscivores)generally results in refocusing of fishing effort on planktivores and a concomitant decline in the average trophic level of the landings. One characteristic of overfished ecosystems is sequential depletion of economic stocks.

Switching between target species occurs when resources of economic importance are markedly reduced in abundance by overfishing and when there are other more abundant stocks available. Piscivores and valuable invertebrate stocks are particularly vulnerable to this fishing pattern. Overfishing and depletion of some stocks may become so severe that they may be regarded as economically extinct.

The symptoms of ecosystem overfishing include: reductions in diversity; reductions in aggregate production of exploitable resources; decline in mean trophic level, increased by-catch; greater variability in abundance of species; greater anthropogenic habitat modification; and, in extreme cases, change to alternative stable species regimes.

Economic Overfishing

A level of fish harvesting that is higher than that of economic efficiency; harvesting more fish than necessary to have maximum profits for the fishery occurs when the costs of fishing effort are greater than the revenue generated from fishing: Incomes of fishermen continuously decline 10 % eper-the spawning stock biomass that will produce 10 per cent of the eggs that would be produced if all females were laft to live their natural life span, with no fishing.

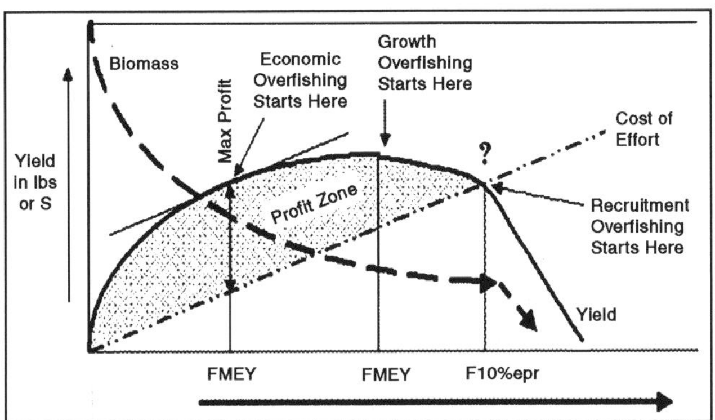

Fig. Fishing Mortality Rate-F or Effective Fishing Effoer(boats, Days, etc.) MEY-maximum Conomice Yield, MSY-Maximum Sustainable Yield

Note that this optimum level of effort is always less than that required to extract maximum sustainable yied (MSY) and that, therefore, maximum economic yield is always less than MSY. The stated or implicit goals of many management schemes; and that beyond MSY, subsidies will reduce catches(by reducing total cost).

- Biological overfishing.
- The combination of growth and recruitment overfishing leading to catch decline on the right, descending side of surplus production models.

Catching such a high proportion of one or all age classes in a fishery as to reduce yields and drive stock biomass, and spawning potential below safe levels. Biological overfishing occurs when fishing levels are higher that those required for extracting the Maximum Sustainable Yield (MSY) of a resource and when recruitment starts to decrease statistically.

Malthusian Overfishing

Desperate attempt to fish to survive where fishers destroy the environment that the fish (and they) depend on for survival using poisons, dynamite, etc. In developing countries, small-scale fishers are usually poor and fishing is an alternative employment opportunities to them. Overtime, the number of these fishers usually increase both because of internal recruitment (*i.e.*, their male children) and through new entrance, *i.e.*, new fishers recruited from other sectors usually to landless farmer to whose fishing becomes an occupation of last resort. Occurs these poor fishers, lacking the usual alternative of traditional fishers are faced with declining catches and induce wholesale resource destruction in their effort to maintain their incomes. This may involve in order of seriousness and generally in temporal sequence such as; Use of fishing techniques, gear and/or mesh sizes not sanctioned by government, use of gears not sanctioned within the fisher communities. Use of gears that destroy the resource base and use of destructive gear such as dynamite and fish poisons that endanger the fishers themselves.

NEW FISHING TECHNOLOGIES AND FISHING METHODS

The national perspective: Following the second world war the rapid expansion of fishing capacity in many developing countries was promoted through the so-called Blue Revolution.

New fishing technologies and fishing methods coupled with a desire to improve nutrition and generate foreign exchange led many countries to build on fishing activity already present. Some of this development was as part of national development programmes, in other cases, international loans and funding programmes helped promote the development of national fishing industries. As a direct result of this drive to build up national fishing capacity, the participation of developing countries in global fishing activity rose dramatically between 1955 and the 1980s. The rewards to be gained from participating in the global fishing market were potentially large, and many benefited, albeit in the short-term, from their forays into fishing.

Increased export-revenues and the multiplier effects associated with expanding industries all helped boost the development of many developing coastal states.

The long-term consequences of such developments, have, however, had a dramatic effect. Rising incomes and increased borrowing of cheap petro-dollars contributed to chronic debt problems as the global economy staggered to a halt in the early 1980s; an attitude of positive speculation on fish-stocks, lack of proper assessment and regulation led to the over-exploitation of a number of important stocks. As a consequence of the international debt burden and the strictures placed on the economy by the Structural Adjustment Programmes, trade-offs between local and national needs put further pressure on resources. The downsizing of the state and the increased role of the market as the resource allocation mechanism saw many developing countries expand their export fisheries to generate revenue whilst at the same time the power to regulate such activity was hampered bythe limited capacity of Fisheries Departments to undertake interventions. The push for export revenue and the renting of grounds to Distant Water Fleets has resulted in spatial conflicts between industrial and artisanal fleets, state support for local fisheries has been constricted and external economic pressures now drive national policy formation.

NATURAL RESOURCES FORM THE SOCIAL SAFETY

From a local perspective: In agricultural-based economies, natural resources form the social safety-net that supports large parts of the population. Access to common property natural resources such as timber, fuelwood, grazing land, fisheries, forest products and irrigation water are fundamental to the livelihoods of many of the world's poor. These natural resources are often held as common property and the survival of rural populations has traditionally relied upon a complex set of institutional arrangements that govern use of and access to these resources. The formation of institutional access and use arrangements is well-documented. Whilst many communities have succeeded in maintaining their common property resources through collective action, in some cases pressures internal and external to the community have caused arrangements to collapse. The outcome of a resource becoming open-access is likely to be over-exploitation and the dissipation of resource rents. In terms of development, then, as the ability of a community of users to exercise rights over their resource is eroded so the role of the resource as a safety-net is undermined.

The pressures that lead to the erosion and collapse of common property resources are diverse. They can result from a rapid rise in population (either through natural growth or an influx of migrants); changes to the ecological status quo (through droughts or flooding, for example); national policy changes that affect rural employment rates; technological change that results in higher harvesting potential and social change that can alter income differentials, systems of property rights and market systems. An important outcome of the breakdown of common property arrangements is the emergence of conflict between stakeholders.

However the precise mechanism is not fully understood. An anatomy of conflict A review of the conflict literature reveals numerous theoretical approaches to describing and explaining conflict. Conflict is a function of social structure (sociology), of power or class relations (political science) and of individual utility maximisation (economics). It is both positive and negative, constructive and destructive, violent, coercive and non-violent. Thisdiversity of approaches reflects the wide range of disciplines that have addressed the subject. There are many possible definitions of conflict, the following synthesis of definitions is used in this paper:

Conflict is a situation of non-cooperation that involves groups of people with differing goals and objectives and yet Conflict is dynamic and can be a positive catalyst for change Although the term 'conflict' often has negative values attached to it, this is not necessarily the case. Conflict can be described as negative when the outcomes are a zero-sum game (no-one benefits) rather than a positive sum-game (everyone benefits) or where there is a deadweight loss of social resources as a result of the 'guns vs butter' argument.

However, it is also important to bear in mind that conflict can be positive and attempts should never be made to eradicate it completely or to prevent it emerging. Conflict encourages goods to be produced more cheaply, government to become more efficient, flaws in the set-up of institutions to be ironed out and allows society to function efficiently by resolving small conflicts often. At this point, the question of the difference between conflict and competition arises. Competition implies the existence of rules, conflict arises when those rules are, for whatever reason, disbanded or ignored. In order to help clarify whether conflict is positive or negative it is often helpful to look at what the conflict is over and how fundamental that disagreement is to the social status quo.

Aubert, Boulding and Powelson all differentiate between conflicts that are 'within consensus' and those that are 'over consensus'. In the former case the parties to the conflict agree about the value of what they seek but not the means of achieving it. In the latter case the parties to the conflict are unable to agree on the value of what they seek nor on how to achieve it. The potential impact of conflict is thus dictated by the degree of consensual framework within which they are contested and the degree of conflict over basic consensus.

AN ECONOMIC EXPLANATION FOR CONFLICT

Economic theory provides two reasons why conflicts emerge over natural resources:

- The allocation of scarce resources requires trade-offs which become increasingly difficult as the demand for and supply of resources changes and
- Short-term personal gain wins out over long-term social needs or benefits.

The resource allocation issue: Natural resource conflicts occur when the resource in question has become so scarce or degraded as to raise issues of allocation amongst the community of users. Under perfect environmental conditions, as the ratio of users to resource grows so expansion takes place: extra land is brought into cultivation, new areas of forest are exploited, different species are fished or fishery activity moves along the coast or further out to sea. Powelson (1972:33) argues that so long as the answer to 'who gets how much' is resolved by producing more, conflicts over allocation are positive because they encourage growth.

When the ecological boundaries of the resource have been reached, further expansion is no longer possible (without adverse consequences) and the division of the metaphorical cake has to change because the option of increasing the size of the cake is no longer possible. At this point, resource allocation mechanisms come under increased pressure to satisfy all users. The equity of resource allocation tends to diminish as the ability to expand production decreases. As government departments attempt to placate

all the stakeholders, so resource allocation decisions have to change and conflict ensues as certain stakeholders gain precedence over others. This is of course nothing new. The trade-offs between equity and efficiency occur in all markets and are faced by all governments. In the developing world however, where democracy may be weak and where powerful interest groups are able to co-opt and unduly influence the decision making process, managing the trade-offs is difficult.

Short-term gain vs social needs: There is a large body of literature that uses game theory and behaviour modelling to explain why common property management institutions (in particular) can fail. The Nash Equilibrium and the Prisoner's Dilemma illustrate this point neatly: in the classic decision making model-the Prisoner's Dilemma-the best possible move for player A given the best possible move for player B is to not co-operate.

Thus the socially optimal outcome of the 'game' is not a Nash Equilibrium notes, in a similar vein that the short term benefits of cooperating are often outweighed by the costs.

Consequently a community opts for the least-cost option which is to withhold their contribution to the collective goal. A third explanation for the emergence of conflict has also arisen from the more recent New Institutional Economics school, this sees institutions and transaction costs as a key element.

From a new institutional economics perspective, institutions exist to minimise transaction costs and are a means of transcending the social welfare dilemmas that arise out of individual action and help maximise collective welfare. They are also the mechanism for dealing with 'gaps' left by market failure in insurance and risk assessment. Transaction costs are literally the costs involved in negotiating a transaction and have been described as the economic equivalent of friction in the world of physics. Transaction costs are the costs associated with gaining information, making decisions and carrying out decisions.

In terms of fisheries, they can be divided into the ex-ante costs of collecting information and making collective decisions in the fishery and the ex-post costs associated with implementing collective decisions. Kuperan et al argue that transaction costs in fisheries arise, mainly, from the fact that the fishery involve multiple stakeholders and differing objectives and long-term goals. Institutions are important as resource allocation mechanisms-markets and fisheries departments, for example, both have a role to play in allocating scarce resources. What is more, it is now recognised that institutions matter for economic development and recent development research has recognised the need to 'get the institutions right' if effective economic development is to be achieved.

Katterman points out that the OECD has finally recognised that technical co-operation, the long-time philosophical foundation of development practice, can amount to nothing if the institutions that it relies upon for success are weak. Thus for both successful natural resource management and economic development institutions able to adapt to changing circumstances are needed. There are two explanations about why institutions might fail in the face of change.

The first relates to the supply and demand for institutional change, the second to transaction costs. Thomson, Feeny and Oakerson and Feeny discuss the supply and demand model for institutional change: when the demand for institutional change (to capture gains not possible under existing arrangements) outstrips the ability to supply change, failure emerges. They list relative factor or product prices, the size of the market, technological change and fundamental decisions of government as causes of change that lead to what they term 'institutional disequilibrium'. However, they also recognise that change depends on the State's willingness and ability to help new institutions emerge. This is a particularly pertinent issue in the developing world where, as discussed before, state capacity may be underdeveloped. Because in the NIE perspective institutions evolve to minimise transaction costs (through minimising uncertainty)-when it is no longer able to do this effectively its position is weakened and it is increasingly unable to deliver services effectively or be allocatively efficient. Using the NIE paradigm and the supply and demand argument above it could, therefore, be argued that a rise in transaction costs, over and above the institution's ability to accommodate this change leads to inefficient operation. This rise in transaction costs could be due to development pressures (political and economic), environmental scarcity (perceived or otherwise) and structural problems (political and economic) which increase the costs mentioned earlier: the costs of collective decision making, information gathering and collective operation.

In the case of fisheries management, strong and flexible institutions are required at both the national and local level. Arguably, the more nested the structure, the lower the transaction costs between the top and the bottom of the system and the more efficient the institution in its allocation role.

In many developing countries this may not be the case. Communication between the different layers of fisheries management are frustrated, the legitimacy of the ruling body to assign resources is often missing or contested and political factions and the action of rent-seeking elites influence the vertical relationship. Institutions, conflicts and fishing in Ghana The political and economic context

Ghana, the former British colony of the Gold Coast, became the first African colonial state to gained independence in 1957. It has had a number of military coups, but is set, in December 2000 to become the first African state in which a military ruler has handed power to a civilian government through the ballot box. Unlike many other countries in the region, Ghana has experienced comparatively little unrest with its neighbours or amongst its diverse tribal and linguistic groups, and it is perhaps this aura of stability and peace at a national level which influences conditions at a local level. In the 43 years since independence Ghana has made some remarkable advances in development and has also suffered some serious set-backs.

Despite significant advances made in improving some of the standard development indicators (life expectancy has risen by 4 years since 1995; adult literacy has risen by 44% since 1970), Ghana is currently struggling with the dual effects of economic adjustment measures and external shocks. Economic reforms have seen public sector budgets slashed and the impact in the natural resources sector, to name but one, has been devastating. All departments are under funded, and in the fisheries department, this has lead, in particular to a reduced ability to monitor fisheries effectively and enforce regulations. The need to raise revenue has seen the introduction of VAT at 10% which is set to rise to 12.5% later this year. Despite well intentioned publicity campaigns to educate the public as to the need for VAT, this is unlikely to ease the blow of a significant price rise on all VATable goods-particularly on basic grains that have shown sharp price rises since last year. Coupled to this is the rapidly devaluing national currency which, while making Ghanaian exports cheaper, makes servicing the national debt (currently $580 million) more expensive daily and dramatically increases the costs of vital industrial inputs. Ghana is an oil importer, and recent OPEC price rises have also hit the economy hard. While exports have grown aided by improving terms of trade and export initiatives put in place as a result of economic reforms, these have tended to concentrate on cash crops such as cocoa.

In the context of fisheries, the prevailing economic conditions are having a number of effects. Firstly the day to day costs of fishing are rising. Fuel prices not only impinge on the length of fishing trips, but exchange-rate depreciation has increased the costs of imported inputs such as nets, ropes and outboard motors. The knock-on effects of such price rises are widespread. Education and health at the local level both suffer as more and more income is diverted away from the 'add-on luxuries' into the basic needs for survival. Although things have not reached crisis level as yet, there is evidence of malnutrition in fishing villages and primary schools lack basic teaching materials (supplied by the community). Despite this catalogue of economic problems, some very interesting results emerged from a recent study of the causes and management of conflict in coastal Ghanaian fishing villages.

ECONOMICS AND MANAGEMENT OF AQUATIC RESOURCES

Between March and May 2000 the Centre for the Economics and Management of Aquatic Resources in collaboration with the Ghanaian Department of Fisheries conducted a survey of 64 fishing villages along the Ghanaian coast. This sample represents 33% of the villages listed in the latest Canoe Frame Survey conducted by the Department of Fisheries. These villages are spread amongst 4 regions. With no information on the variability of the population it was decided to select one third of all the villages in each region on a random basis. In order to gain a complete picture of conflicts within the coastal zone, interviews were also conducted with inshore-vessel owners and a number of industrial fleet operatives.

Ghanaian artisanal canoe communities are organised around the Chief Fisherman. A complex set of procedures has to be undertaken before any interviewing can take place and no interviewing can be conducted with permission from the Chief Fisherman. Consequently we targeted the Chief Fishermen who on most occasions was accompanied by a number of other village elders, usually comprising his right-hand man and the most skilful or successful fisherman in the village.

On a number of occasions large groups of women traders and other fishermen were present at the meeting. For reasons of simplicity and speed a number of tools were adapted from the extensive PRA 'toolbox' to help build PISCES (Participatory Institutional Survey and Conflict Evaluation Study) developed specifically for use with artisanal fishing communities to gather information on institutional arrangements.

Oasis of Calm in a Sea of Troubles

Theory would suggest that given the economic context within which Ghanaian fisheries are operating, the open-access nature of the resource and the inherent instability of pelagic stocks, conflict should be widespread and the fisheries management institutions under considerable pressure. Initial analysis, however, suggests that this is not the case. At the time of writing data collection was still continuing in Central and Western Regions. The summary results below are therefore preliminary results on regions already completed.

Villages as institutions. From the results so far, Ghanaian coastal villages appear to be stable and differ little in their organisational structure along the entire length of the coast. The village chief is at the top of the hierarchy, the Chief Fishermen, first point of call for all fisheries matters sits beneath the Chief. Although by law the Ghanaian coastal line is open access (to Ghanaian citizens), in practice, a complex set of local laws and a system of reciprocity and responsibility govern inshore waters. 91.7% of villages questioned confirmed that although anyone could fish off the beach, in practice, they must seek permission or announce their intentions to the Chief Fisherman, this includes residents and migrants. The remaining 8.3% stated that anyone could fish off the beach, and did not qualify their answer. No one reported ever having a request to fish denied.

Once permission has been granted by the Chief Fisherman he then becomes responsible for the welfare of the fisherman and his kin. This system of 'responsibility' is not unique to Ghana and acts as a kind of insurance scheme. There are many internal migrant fishermen in Ghana-some are migrants that seasonally move from one area to another, others are migrants that have been settled in a village for generations but, because they are from a different language group, are still classed as migrants. The distinction between natives and migrants to so powerful than in one village in Western Region, there are two separate landing sites: the Fante-line and the Ewe-line highlighting the two distinct groups that have fished from the village for many years.

INSTITUTIONALISATION OF FISHERIES MANAGEMENT

The institutionalisation of fisheries management. The local institutional structure of Ghanaian fisheries has, on the face of it, changed very little over the past centuries, and it is perhaps this stability and consistency that helps it remain comparatively peaceful. However, economic reforms, as described above have caused many significant negative changes to the national organisation of fishing in the past five years but in some areas have also ironically tended to bolster the local structures rather than threaten them. The most significant positive change has been the introduction of the Community Based Fisheries Management Programme (CBFM) under the auspices of the Fisheries Sub-sector Capacity Building Project. The success of the CBFM would appear to be largely due to the fact that rather than radically changing the status quo, it has strengthened the existing structures.

The Fisheries Sub-sector Capacity Building Project is a joint venture between the Government of Ghana and IDA/World Bank. Started in October 1995 its main objectives were to improve the long-term sustainability of Ghanaian fisheries. Within this main objective was also the aim to improve the capacity of the Department of Fisheries, address the issue of lack of an active management regime, weak institutional and legal frameworks for fisheries and a growing financial and resource crisis in the industry. In order for the sustainability of Ghanaian fisheries to be improved management plans were needed and for these to be successful they would need the full approval of local communities. The Project will run until 2001 when the funding will finish and the Government of Ghana will be responsible for seeing its continuation.

As part of the CBFM, committees have been formed in each community and their first task was to draw up a list of by-laws governing fishing activity there. By drawing up their own by-laws and submitting them to the District Assembly for ratification, local norms would be lent weight and legitimacy. This has a two-way function. Firstly the community feels that its laws are valued because they are recognised by a higher authority and secondly, it engenders a sense of trust and liaison between the communities and the District Assemblies. The make-up of the committees varies but as a rule consists of the chief fisherman and representatives of the various stakeholders. Typically this would include the indigenous and migrant fishermen, the fish processors, fishmongers and fishtraders, vessels, gear and engine owners, and a representative from the Fishermen's Service Centre.

During the course of the field work the villages in a number of districts in Central Region were going through the process of ratifying the by-laws which had recently been gazetted. The drawing up of by-laws has allowed existing norms to be institutionalised and has allowed 'best-practice' in fishing to be argued out amongst the fishermen and agreed up in a formal setting. A good example of this is the by-law that bans children from the beach during school hours (recognising that schooling is a very important part of village development) and the by-law that formally bans fishing with dynamite (an extremely contentious issue up to this point).

Conflicts and conflict management. In the regions so far covered, violent conflict between fishermen was very rarely mentioned. Day-to-day squabbles and difficulties are, however, common. Interestingly, the pricing of fish is regarded as a conflict, although it could be argued that this is an essential and integral part of how markets work and is a sign of competition not conflict. However, because the ethos of the survey was for the communities to identify what they considered to be conflicts, we have let it stand. Two other conflicts often attracted heated debate: the incursion of semi-industrial trawlers into water less than 30 metres deep (reported by 25% of villages) and fights among women over access to catches and credit facilities (reported by 29.2% of villages). Although there was scant reference to a rising number of canoes or fishermen related to the presence of conflict, two examples stand out in particular.

One was a fishing community in Accra which suffers from its proximity to the city centre and all the urban poverty problems associated with it. Here the Secretary to the Chief Fisherman mentioned that the landing beach was no longer able to cope with the number of canoes and there were many conflicts over landing canoes and off-loading catches. The other was Mumford, a well documented fishing village also reported problems with lack of landing facilities exacerbated by the increased number of canoes present in the village. Based on the answers recorded, conflict in Ghanaian coastal artisanal fisheries can be divided into a number of categories: conflict that results from outside influences, conflict that results from the internal allocation of resources and that which could probably be better described as competition. The typology below attempts to categorise the most frequent problems encountered, it should be noted that at the margins some conflicts could happily sit in a number of boxes. Those conflicts caused by outside influences are the hardest to solve because they involve elements beyond the immediate control of the village.

Because violent conflict is almost unheard of, it is perhaps useful to establish a conflict scale for Ghana that better distinguishes between the various degrees of day-to-day squabbles.

Conflict management at the village level is highly structured and organised and it is perhaps a reflection of this that conflicts are few and far between. All villages in the survey reported that any conflict between fishermen is reported first to the Chief Fisherman who then, along with his panel of elders, comes to a decision on the issue. In the case of damage to nets or boats caused by other canoe owners, the culprit is usually made to pay 2/3 of the damages (recognising thus that fishing is a dangerous activity and incidents are rarely deliberate). Although these cases may take a while to resolve-establishing fault and liability-they are not considered to be onerous. The conflicts that do take longer to resolve are those between different types of vessel-trawlers damaging the nets of canoe fishermen cause a frequent problem.

Even when there are a number of witnesses and broad daylight has enabled the collection of the name and registration details of the offending vessel, the cases can take months to resolve. Such cases are reported to the relevant local fisheries office that then deals with them. Trawler owners often deny all responsibility for the damage and communication and transport difficulties between the administrative centres and the villages further lengthens the process.

During discussions other factors arose that, whilst not identified as conflicts by the villagers, were identified as making life in the villages more difficult. Two examples of these are the incidence of erosion and the rising costs of inputs. Sea erosion was frequently mentioned in villages in Greater Accra and Volta region to the east of Tema harbour. Sea erosion in the Bight of Benin is serious; current large scale project funded by international agencies are working to building sea defences to protect the coastline. Local lore suggests that without these sea defences Keta Lagoon (a significant water body close to the Togo border) will disappear into the sea in five years time. The building of Tema harbour was the most frequently cited reason for erosion in the east of the country. The knock-on effects of the Akosombo Dam, built on the Volta River are also cited for disrupting flooding patterns in the river delta (this has resulted in decreasing water levels in lagoons) and the disappearance of certain fish species at the mouth of the river. The rising price of inputs was universally mentioned as one of the hardest issues to contend with. Given that changing economic circumstances affect how communities deal with the day to day running of their lives, it is hardly surprising that there is a perceived link between increased economic hardship and conflict.

4

Method of Fishing Techniques

Fishing techniques are methods for catching fish. The term may also be applied to methods for catching other aquatic animals such as molluscs (shellfish, squid, octopus) and edible marine invertebrates. Fishing techniques include hand gathering, spearfishing, netting, angling and trapping. Recreational, commercial and artisanal fishers use different techniques, and also, sometimes, the same techniques.

Recreational fishers fish for pleasure or sport, while commercial fishers fish for profit. Artisanal fishers use traditional, low-tech methods, for survival in third-world countries, and as a cultural heritage in other countries. Mostly, recreational fishers use angling methods and commercial fishers use netting methods. There is an intricate link between various fishing techniques and knowledge about the fish and their behaviour including migration, foraging and habitat. The effective use of fishing techniques often depends on this additional knowledge. Which techniques are appropriate is dictated mainly by the target species and by its habitat.

POSSIBLE OF HAND FISHING

It is possible to fish and gather many sea foods with minimal equipment by using the hands. Gathering seafood by hand can be as easily as picking shellfish or kelp up off the beach, or doing some digging for clams or crabs. The earliest evidence for shellfish gathering dates back to a 300,000 year old site in France called Terra Amata. This is a hominid site as modern Homo sapiens did not appear until around 50,000 years ago.

- Flounder tramping-Every August, the small Scottish village of Palnackie hosts the world flounder tramping championships where flounder are captured by stepping on them.
- Noodling: is practiced in the United States. The noodler places his hand inside a catfish hole. If all goes as planned, the catfish swims forward and latches onto the noodler's hand, and can then be dragged out of the hole.
- Pearl divers-traditionally harvested oysters by free-diving to depths

of thirty metres. Today, free-diving recreational fishers catch lobster and abalone by hand.

- Trout binning-is another method of taking trout. Rocks in a rocky stream are struck with a sledgehammer. The force of the blow stuns the fish.
- Trout tickling-In the British Isles, the practice of catching trout by hand is known as trout tickling; it is an art mentioned several times in the plays of Shakespeare.

Fishing of Netting

Fishing nets are meshes usually formed by knotting a relatively thin thread. About 180 AD the Greek author Oppian wrote the *Halieutica*, a didactic poem about fishing. He described various means of fishing including the use of nets cast from boats, scoop nets held open by a hoop, and various traps "which work while their masters sleep". Netting is the principal method of commercial fishing, though longlining, trolling, dredging and traps are also used.

- Artisanal techniques
- Chinese fishing nets-are shore operated lift nets. Huge mechanical contrivances hold out horizontal nets with diameters of twenty metres or more. The nets are dipped into the water and raised again, but otherwise cannot be moved.
- Lampuki nets-are an example of a traditional artisanal use of nets. Since Roman times, Maltese fishers have cut the larger, lower fronds from palm trees which they then weave into large flat rafts. The rafts are pulled out to sea by a luzzu, a small traditional fishing boat. In the middle of the day, lampuki fish (the Maltese name for mahi-mahi) school underneath the rafts, seeking the shade, and are caught by the fishers using large mesh nets.
- Cast nets-are round nets with small weights distributed around the edge. They also called *throw nets.* The net is caste or thrown by hand in such a manner that it spreads out on the water and sinks. Fish are caught as the net is hauled back in. This simple device has been in use, with various modifications, for thousands of years.
- Drift net-are nets which are not anchored. They are usually gillnets, and are commonly used in the coastal waters of many countries. Their use on the high seas is prohibited, but still occurs.
- Ghost nets-are nets that have been lost at sea. They can be a menace to marine life for many years.
- Gillnets-catch fish which try to pass through by snagging on the gill covers. Trapped, the fish can neither advance through the net nor retreat.
- Hand nets-are small nets held open by a hoop. They have been

used since antiquity. They are also called scoop nets, and are used for scooping up fish near the surface of the water. They may or may not have a handle–if they have a long handle they are called *dip nets*. When used by anglers to help land fish they are called *landing nets*. Because hand netting is not destructive to fish, hand nets are used for tag and release, or capturing aquarium fish.

- Seine nets-are large fishing nets that can be arranged in different ways. In purse seining fishing the net hangs vertically in the water by attaching weights along the bottom edge and floats along the top. Danish seining is a method which has some similarities with trawling. A simple and commonly used fishing technique is beach seining, where the seine net is operated from the shore.
- Surrounding net-
- Trawl nets-are large nets, conical in shape, designed to be towed in the sea or along the sea bottom. The trawl is pulled through the water by one or more boats, called trawlers. The activity of pulling the trawl through the water is called trawling.

METHODS OF SPEARFISHING FISHING

Spearfishing is an ancient method of fishing conducted with an ordinary spear or a specialised variant such as a harpoon, trident, arrow or eel spear. Some fishing spears use slings (or rubber loops) to propel the spear.

- Bowfishing-uses a bow and arrow to kill fish in shallow water from above.
- Gigging-uses small trident type spears with long handles for gigging bullfrogs with a bright light at night, or for gigging suckers and other rough fish in shallow water. Gigging is popular in the American South and Midwest.
- Hawaiian slings-have a sling separate from the spear, in the manner of an underwater bow and arrow.
- Harpoons-Spearfishing with barbed poles was widespread in palaeolithic times. Cosquer cave in Southern France contains cave art over 16,000 years old, including drawings of seals which appear to have been harpooned.
- Polespears-have a sling attached to the spear.
- Modern spearguns-traditional spearfishing is restricted to shallow waters, but the development of the speargun has made the method much more efficient. With practice, divers are able to hold their breath for up to four minutes and sometimes longer. Of course, a diver with underwater breathing equipment can dive for much longer periods.
- Tridents-are three-pronged spears. They are also called leisters or

gigs. They are used for spear fishing and were formerly also a military weapon. They feature widely in early mythology and history.

Methods of Angling and Line Fishing

Angling is a method of fishing by means of an "angle" (hook). The hook is usually attached to a line, and is sometimes weighed down by a sinker so it sinks in the water. This is the classic "hook, line and sinker" arrangement, used in angling since prehistoric times. The hook is usually baited with lures or bait fish. Additional arrangements include the use of a fishing rod, which can be fitted with a reel, and functions as a delivery mechanism for casting the line. Other delivery methods for projecting the line include fishing kites and cannons, kontiki rafts and remote controlled devices. Floats can can also be used to help set the line or function as bite indicators. The hook can be dressed with lures or bait. Angling is the principal method of sport fishing, but commercial fisheries also use angling methods involving multiple hooks, such as longlining or commercial trolling.

Line fishing is fishing with a fishing line. A fishing line is any cord made for fishing. Important parameters of a fishing line are its length, material, and weight (thicker, sturdier lines are more visible to fish). Factors that may determine what line an angler chooses for a given fishing environment include breaking strength, knot strength, UV resistance, castability, limpness, stretch, abrasion resistance, and visibility. Modern fishing lines are usually made from artificial substances. The most common type is *monofilament*, made of a single strand. There are also *braided fishing lines* and thermally fused *superlines*.

- Droplining-a dropline consists of a long fishing line set vertically down into the water, with a series of baited hooks Droplines have a weight at the bottom and a float at the top. They are not usually as long as longlines and have fewer hooks.
- Handlining-is fishing with a single fishing line, baited with lures or bait fish, which is held in the hands. Handlining can be done from boats or from the shore. It is used mainly to catch groundfish and squid, but smaller pelagic fish can also be caught.
- Longlining-is a commercial technique that uses a long heavy fishing line with a series of hundreds or even thousands of baited hooks hanging from the main line by means of branch lines called "snoods". Longlines are usually operated from specialised boats called longliners. They use a special winch to haul in the line, and can operate in deeper waters targeting pelagic species such as swordfish, tuna, halibut and sablefish.
- Slabbing-involves repetitively lifting and dropping a flat lure, usually made of 1 to 2.5 oz of lead painted to look like a baitfish, through a school of actively feeding fish that the angler has located on a fish

finder. Used on white and striped bass in the reservoirs of the southern USA.

- Trolling-is fishing with one or more baited lines which are drawn through the water. This may be done by pulling the line behind a slow moving boat, or by slowly winding the line in when fishing from the land. Trolling is used to catch pelagic fish such as mackerel and kingfish.
- Trotlining-a trotline is like a dropline, except that a dropline has a series of hooks suspended vertically in the water, while a trotline has a series of hooks suspended horizontally in the water. Trotlines can be physically set in many ways, such as tying each end to something fixed, and adjusting the set of the rest of the line with weights and floats. They are used for catching crabs or fish, such as catfish, particularly across rivers.

Angling with a Rod

Fishing rods give more control of the fishing line. The rod is usually fitted with a fishing reel which functions as a mechanism for storing, retrieving and paying out the line. Floats may also be used, and can function as bite indicators. The hook can be dressed with lures or bait.

- Bank fishing-fishing from river banks and similar shorelines. Bank fishing is usually performed with a rod and reel, alhough nets, traps, and spears can also be used. People who fish from a boat can sometimes access more areas in prime locations with greater ease than bank fishermen. However many people don't own boats and find fishing from the bank has its own advantages. Bank fishing has its own requirements, and many things come into play for success, such as local knowledge, water depth, bank structure, location, time of day, and the type of bait and lures.
- Casting-the act of throwing the fishing line out over the water using a flexible fishing rod. The usual technique is for the angler to quickly flick the rod from behind towards the water. Casting is also a sport adjunct to fishing, much as shooting is to hunting. The sport is supervised by the International Casting Sport Federation, which sponsors tournaments and recognizes world records for accuracy and distance.
- Float tubes-small doughnut-shaped boats with an underwater seat in the "hole". Float tubes are used for fly fishing and enable the angler to reach deeper water without splashing and disturbing stillwater fish.
- Fly fishing-the use of artificial flies as lures. These are cast with specially constructed fly rods and fly lines. The fly line (today, almost always coated with plastic) is heavy enough cast in order to send the

fly to the target. Artificial flies vary dramatically in size, weight and colour. Fly fishing is a distinct and ancient angling method, most renowned as a method for catching trout and salmon, but employed today for a wide variety of species including pike, bass, panfish, and carp, as well as marine species, such as redfish, snook, tarpon, bonefish and striped bass. There is a growing population of anglers whose aim is to catch as many different species as possible with the fly.

- Rock fishing-fishing from rocky outcrops into the sea. It is a popular pastime in Australia and New Zealand. It can be a dangerous pastime and claims many lives each year.
- Surfcasting-fishing from a shoreline using a rod to cast into the surf. With few exceptions, surf fishing is done in saltwater, often from a beach. The basic idea of most surfcasting is to cast a bait or lure as far out into the water as is necessary to reach the target fish from the shore. This may or may not require long casting distances and muscular techniques. Basic surf fishing can be done with a surfcasting rod between seven and twelve feet long, with an extended butt section, equipped with an appropriate spinning or conventional casting reel. Dedicated surfcasters usually possess an array of terminal and other tackle, with rods and reels of different lengths and actions, and lures and baits of different weights and capabilities. Depending on fishing conditions and the fish they are targeting, such surfcasters tailor bait and terminal tackle to rod and reel and the size and species of the fish. Reels and other equipment need to be constructed so they resist the corrosive and abrasive effects of salt and sand.

Trapping

Traps are culturally almost universal and seem to have been independently invented many times. There are essentially two types of trap, a permanent or semipermanent structure placed in a river or tidal area and pot-traps that are baited to attract prey and periodically lifted.

- Dam fishing-An artisanal technique called dam fishing is used by the Baka pygmies. This involves the construction of a temporary dam resulting in a drop in the water levels downstream.
- Basket weir fish traps-were widely used in ancient times. They are shown in medieval illustrations and surviving examples have been found. Basket weirs are about 2 m long and comprise two wicker cones, one inside the other — easy to get into and hard to get out.
- Fishing weir-In medieval Europe, large fishing weir structures were constructed from wood posts and wattle fences. 'V' shaped structures in rivers could be as long as 60 metres and worked by directing fish

towards fish traps or nets. Such fish traps were evidently controversial in medieval England. The Magna Carta includes a clause requiring that they be removed: *"All fish-weirs shall be removed from the Thames, the Medway, and throughout the whole of England, except on the sea coast".*

- Fish wheels-operate alongside streams, much as a water-powered mill wheel. A wheel complete with baskets and paddles is attached to a floating dock. The wheel rotates due to the current of the stream. The baskets on the wheel capture fish travelling upstream and transfer them into a holding tank. When the holding tank is full, the fish are removed.
- Lobster traps-also called lobster pots, are traps used to catch lobsters. They resemble fish traps, yet are usually smaller and consist of several sections. Lobster traps are also used to catch other crustaceans, such as crabs and crayfish. They can be constructed in various shapes, but the design strategy is to make the entry into the trap much easier than exit. The pots are baited and lowered into the water and checked frequently. Historically lobster pots were constructed with wood or metal. Today most traps are made from checkered wire and mesh. It is common for the trap to be weighted down with bricks. A bait bag is hung in the middle of the trap. In theory the lobster walks up the mesh and then falls into the wire trap. Bait varies from captain to captain but it is common to use herring. In commercial lobstering five to ten of these traps will be connected with line. A buoy marks each end of the string of pots. Two buoys are important to make retrieval easier and so captains don't set their traps over each other. Each buoy is painted differently so the various captains can identify their traps.

Other Angling

- Bottom fishing-is fishing the bottom of a body of water. In the United Kingdom it is called "legering". A common rig for fishing on the bottom is a weight tied to the end of the line, with a hook about an inch up line from the weight. The method can be used both with hand lines and rods. There are fishing rods specialized for bottom fishing, called "donkas". The weight is used to cast or throw the line an appropriate distance. Bottom fishing can be done both from boats and from the land. It targets groundfish such as sucker fish, bream, catfish, and crappie.
- Fishing cannon-
- Ice fishing-is the practice of catching fish with lines and hooks through an opening in the ice on a frozen body of water. It is practised by

hunter-gatherers such as the Inuit and by anglers in other cold or continental climates.

- Kayak fishing-has a long history, and has gained popularity in recent times. Many of the techniques used are the same as those used on other fishing boats, apart from difference is in the setup, how each piece of equipment is fitted to the kayak, and how each activity is carried out on such a small craft.
- Kite fishing is said to have been invented in China. It was, and still is, used by the people of New Guinea and other Pacific Islands-either by cultural diffusion from China or independent invention. Kites can provide the boatless fishermen access to waters that would otherwise be available only to boats. Similarly, for boat owners, kites provide a way to fish in areas where it is not safe to navigate such as shallows or coral reefs where fish may be plentiful. Kites can also be used for trolling a lure through the water. Suitable kites may be of very simple construction. Those of Tobi Island are a large leaf stiffened by the ribs of the fronds of the coconut palm. The fishing line may be made from coconut fibre and the lure made from spiders webs. Modern kitefishing is popular in New Zealand, where large delta kites of synthetic materials are used to fish from beaches, taking a line and hooks far out past the breakers. Kite fishing is also emerging in Melbourne where sled kites are becoming popular, both off beaches and off boats and in freshwater areas. The disabled community are increasingly using the kites for fishing as they allow mobility impaired people to cast the bait further out than they would otherwise be able to.
- Boat anglers-Fishing is usually done either from a boat or from a shoreline or river bank. When fishing from a boat, pretty much any fishing technique can be used, from nets to fish traps, but some form of angling is by far the most common. Compared to fishing from the land, fishing from a boat allows more access to different fishing grounds and different species of fish. Some tackle is specialised for boat anglers, such as sea rods.
- Remote control fishing-Fishing can also be done using a remote controlled boat. The boat is usually one to three feet long and runs on a small DC battery. A radio transmitter controls the boat. The fisherman connects the fishing line/bait to the boat; drives it; navigating the water by manipulating the remote controller. The technique is growing in popularity. The methods of remote control vary, from tying a piece of fishing line and bait to the boat, and driving, (retrieving) the hooked fish, with the boat. Another method is using an apparatus designed so that once the fish is hooked the line

disconnects from the boat, and the fish is retrieved with the fishing rod.

Animals

- Cooperative human-dolphin fisheries date back to the ancient Roman author and natural philosopher Pliny the Elder. A modern human-dolphin fishery still takes place in Laguna, Santa Catarina, Brazil. Here, dolphins drive fish towards fishermen waiting along the shore and give them a signal when they can cast their nets. The dolphins then feed off the fish that manage to escape the nets.
- Cormorant fishing-In China and Japan, the practice of cormorant fishing is thought to date back some 1300 years. Fishermen use the natural fish-hunting instincts of the cormorants to catch fish, but a metal ring placed round the bird's neck prevents large, valuable fish from being swallowed. The fish are instead collected by the fisherman.
- Frigatebirds fishing-The people of Nauru used trained frigatebirds to fish on reefs.
- Portuguese Water Dogs-Dating from the 1500s in Portugal, Portuguese Water Dogs were used by fishermen to send messages between boats, to retrieve fish and articles from the water, and to guard the fishing boats. Labrador Retrievers have been used by fishermen to assist in bringing nets to shore; the dog would grab the floating corks on the ends of the nets and pull them to shore.
- Remora fishing-The practice of tethering a remora, a sucking fish, to a fishing line and using the remora to capture sea turtles probably originated in the Indian Ocean. The earliest surviving records of the practice are Peter Martyr d'Anghera's 1511 accounts of the second voyage of Columbus to the New World. However, these accounts are probably apocryphal, and based on earlier, no longer extant accounts from the Indian Ocean region.

Other Techniques

- Artisanal techniques
- Basnig-
- Electrofishing-is another recently developed technique, primarily used in freshwater by fisheries scientists. Electrofishing uses electricity to stun fish so they can be caught. It is commonly used in scientific surveys, sampling fish populations for abundance, density, and species composition. When performed correctly, electrofishing results in no permanent harm to fish, which return to their natural state a few minutes after being stunned.
- Fish aggregating devices-are man-made objects used to attract pelagic

fish such as marlin, tuna and mahi-mahi (dolphin fish). They usually consist of buoys or floats tethered to the ocean floor with concrete blocks.

- Dredging-There are types of dredges used for collecting scallops, oysters or sea cucumbers from the seabed. They have the form of a scoop made of chain mesh and they are towed by a fishing boat. Dredging can be destructive to the seabed, because the marine life is unable to survive the weight of the dredge. It is extremely detrimental to coral beds since they take centuries to rebuild themselves. Unmonitored dredging can be compared to unmonitored forest clearing, where it can wipe out ecosystems. Nowadays, this method of fishing is often replaced by mariculture or by scuba diving.
- Fish finders-are electronic sonar devices which indicate the presence of fish and fish schools. They are widely used by recreational fishermen. Commercially, they are used with other electronic locating and positioning devices.
- Fishing light attractors-use lights attached to some structure to attract fish and bait fish. Fishing light attractor are operated every night. After a while, fish discover the increased concentration of bait surrounding the light. Once located, the fish return regularly, and can be harvested.
- Flossing-
- Harvesting machines-have recently been developed for commercial fishing. Harvesting machines use pumps to pump fish out of the sea. Dredges have also been mechanized so that they directly transfer mollusks to the surface as are dredged.
- Payaos-a type of fish aggregating device used in Southeast Asia, particularly in the Philippines. Payaos were traditionally bamboo rafts for handline fishing before World War II, but modern steel payaos use fish lights and fish location sonar to increase yields. While payaos fishing is sustainable on a small scale, the large scale, modern applications have been linked to adverse impacts on fish stocks.
- Shrimp baiting-is a method used by recreational fisherman for of catching shrimp. It uses a cast net, bait and long poles. The poles are used to mark a specific location and then bait is thrown in the water near the pole. After several minutes the cast net is thrown as close to the bait as possible and shrimp are caught in the net. In the 1980s the sport became popular in the south eastern coastal states of the USA.

DESTRUCTIVE TECHNIQUES

Destructive fishing practices are practices that easily result in irreversible damage to aquatic habitats and ecosystems. Many fishing techniques can be destructive if used inappropriately, but some practices are particularly likely to result in irreversible damage. These practices are mostly, though not always, illegal. Where they are illegal, they are often inadequately enforced. Some examples are:

- Explosives-Dynamite or blast fishing, is done easily and cheaply with dynamite or homemade bombs made from locally available materials. Fish are killed by the shock from the blast and are then skimmed from the surface or collected from the bottom. The explosions indiscriminately kill large numbers of fish and other marine organisms in the vicinity and can damage or destroy the physical environment. Explosions are particularly harmful to coral reefs. Blast fishing is also illegal in many waterways around the world.
- Bottom trawling-
- Cyanide fishing-Cyanides are used to capture live fish near coral reefs for the aquarium and seafood market. This illegal fishing occurs mainly in or near the Philippines, Indonesia, and the Caribbean to supply the 2 million marine aquarium owners in the world. Many fish caught in this fashion die either immediately or in shipping. Those that survive often die from shock or from massive digestive damage. The high concentrations of cyanide on reefs harvested in this fashion damages the coral polyps and has also resulted in cases of cyanide poisoning among local fishermen and their families.
- Fish toxins-
- Muroami-is a destructive artisan fishing method employed on coral reefs in Southeast Asia. An encircling net is used with pounding devices, such as large stones fitted on ropes that are pounded onto the coral reefs. They can also consist of large heavy blocks of cement suspended above the sea by a crane fitted to the vessel. The pounding devices are repeatedly lowered into the area encircled by the net, smashing the coral into small fragments in order to scare the fish out of their coral refuges. The "crushing" effect on the coral heads has been described as having longlasting and practically totally destructive effects.

Ancient remains of spears, hooks and fishnet have been found in ruins of the Stone Age. The people of the early civilization drew pictures of nets and fishing lines in their arts. Early hooks were made from the upper bills of eagles and from bones, shells, horns and plant thorns. Spears were tipped with the same materials, or sometimes with flints. Lines and nets were made from leaves, plant stalk and cocoon silk. Ancient fishing nets were rough in design and material but they were amazingly, as if some now use. Literature on the indigenous fishing practices is very scanty. Baines documented traditional fisheries in the Solomon Island. Use of the herbal fish poisons in catching fishes from fresh water and sea documented from New Caledonia. John documented fishing techniques and overall life style of the Mukkuvar fishing Community of Kanyakumari district of Tamilnadu, India. Tribal people using various plants for medicinal and various purposes extends the use notion for herbal fish stupefying plants. Use of the fish poisons is very old practice in the history of human kind. In 1212 AD King Frederick II prohibited the use of certain plant piscicides, and by the fifteenth century similar laws had been decreed in other European countries as well. All over the globe, indigenous people use various fish poisons to kill the fishes, documented in America and among Tarahumara Indian.

5

Traditional Fishing Arts and Fishing Gear

Traditional fishing arts have been developed over the years to adapt to local conditions (such as the type of coast and nearshore area), the species of fish desired, and the size targeted. The most successful fishing methods of a given region are those that have stood the test of time. Some of the traditional fishing methods used around the world and consider their advantages and disadvantages. Each method shows a continuum of development with evolution resulting from modernizing factors. Traditional fishing arts in various stages of modernization could be transferred and applied in new regions with the technical level appropriate for the local conditions. The adaptation of new technologies could help small-scale fisheries increase their catch. They could compete more effectively with industrial fisheries or exploit a previously unexploited resource. Energy-efficient technologies are recommended where possible.

The introduction of any new fishing technology always demands good national management and regulation. Vessels must also be matched with new methods or gear. As gear becomes more complex, it may require upgrading of vessels in size, power, and design. The site specificity of fishing arts should always be considered.

PASSIVE GEAR

Passive gear is stationary. It does not have to be dragged or towed to capture fish. Long lines, traps, weirs, and gill nets effectively fish by themselves. The catch is recovered by simply removing the gear from the water after a period of time.

HOOKS AND LINES

The simplest form of fishing requires only a line and a baited hook. The line is cast into the water where the fish supposedly are, the fish take the bait and are hauled in. Lines may be cast by ingenious methods. In Oceania, the line is wound around a stone and thrown from the shore into the water. Hook and line fishing is inexpensive and easy. Almost any boat or shoreline can be used and the catch is live and of high quality. A wide variety of sizes and types of hooks and lures can be used, allowing very selective fishing. Tuna fishing with poles and lines continues to be widely practiced and productive. In spite of these advantages, line fishing is labor intensive. A very limited number of fish can be captured per line and usually some type of bait is required. Line-fishing methods can be made more efficient if multiple hooks on a line are used. Often these are attached in pairs to form balanced lines. A single, branched rod, used in Lake Tanganyika fisheries, also allows one person to fish an increased number of lines and hooks. However, the number of lines that one person can hold is limited.

SET LINES

The use of set lines can increase the number of lines deployed without requiring the constant presence of the fisherman. Such lines must be checked regularly because predators will devour any fish caught if the lines are not promptly recovered. Fishing rods can be set untended in shallow waters or on the beach. In the ocean, set lines may be suspended from the surface.

LONGLINES

Longlines are unwatched lines with multiple hooks. They can be used at the surface, suspended in the water column, or fixed on or near the bottom. Japanese and Italian fishermen use sailing rafts to tow longlines away from their boats. Longlines may be set from the beach by means of sailing rafts or kites if winds are favorable. Surface longlines are used to capture tuna, shark, and billfish. Subsurface and bottom-set longlines are used to catch cod, grouper, snapper, drum, bream, halibut, haddock, hake, and flatfish.

Fig. Multiple-baited hooks on a line can increase the catch.

Fig. Long lines of baited hooks can be deployed at the surface, midwater, at the bottom, or vertically in the water column.

An alternative to bottom-set longlines is a vertical fish stick. This device is hung from a surface float just off the bottom. It has rigid branches to allow multiple hooks without snagging. Fishermen can use local materials to fabricate this gear. Hook-and-line fishing methods offer a number of advantages. They involve low capital and energy investments and labor-intensive operations. Species and size can be selected by the position of the hook in the water column, the hook size, and by the bait type and size. Small-scale fisheries using only open boats can easily adopt hook-and-line methods.

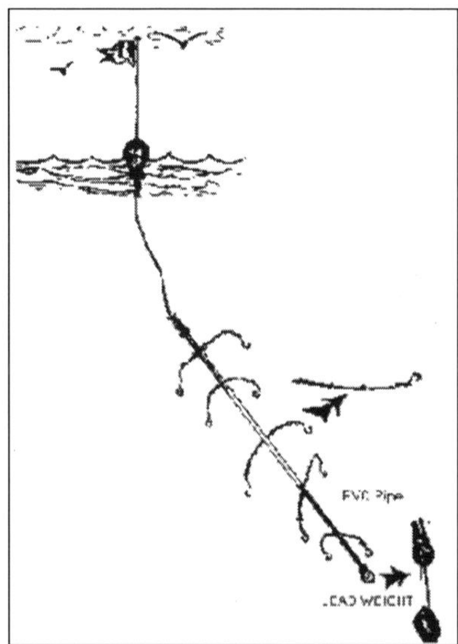

Fig. This 2.5-m rod has 5 rigid cross branches, each with 2 hooks. Adjacent cross branches, are set at about 90 gr. To each other for greater spacing between hooks.

At the same time, the hooks generally require bait (which may be expensive) and baiting is time consuming. It may be difficult to store longlines and their catch on a small vessel. Moreover, a high degree of skill is involved in deploying and retrieving longlines, unless expensive mechanized equipment is used. Modernization in longline fisheries generally involves the mechanization of hauling. If available, hydraulic or electrical drives offer better control, lower maintenance, and variable power.

Simple mechanical hauling techniques can increase the range and depth of hand-line fishing. Manually operated reels can be constructed of local materials and used to reduce the effort needed to handle gear. A vehicle wheel (from a bicycle or automombile) can be fitted with a handle and mounted on the boat to create a simple roller, which facilitates handling lines or nets. A simple wooden hand hauler can also be constructed. The placement, design and dimensions of the hauler can vary according to the size and length of line, and the type and depth of fishing.

TRAPS AND POTS

Traps are devices that fish or shellfish enter in search of shelter or food, or because an obstacle is placed in the fish's normal path of migration. They are designed so that getting out is harder than getting in. Traditional techniques employing traps and pots have developed in all regions of the world to catch demersal species Their design and operation match the specific conditions and be havior of fish in a given area. Hence the importance of local knowledge in design and placement cannot be overemphasized.

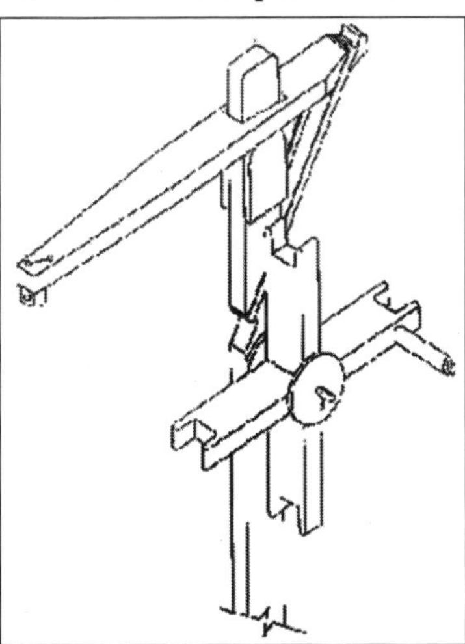

Fig. A simple wooden handreel can ease landing larger fish. (D. C. Cook).

Traps and pots can be quite specifically tailored to species and size. They may be constructed of local materials, generally at low cost, and usually require no bait. An additional advantage of this fishing method is the high quality of the live catch. At the same time, their construction requires skill and knowledge of fishing conditions. Traps and pots are bulky and can occupy considerable space on a vessel. Handling is difficult, and manual hauling is arduous. A major disadvantage of this technique is the high loss of gear due to theft, storm damage, degradation of materials, and the inability to locate the gear once it is deployed.

A lost trap may continue to operate ("ghost fishing"), depleting marine resources. Traps increasingly tend to be constructed of more durable materials, such as wire mesh and vinyl-clad wire netting, instead of traditional woods and fibers. Electrical, mechanical, and hydraulic hauling techniques are also modernizing this fishing system. There are thousands of trap designs throughout the world, many of which would be innovative outside of their region of use.

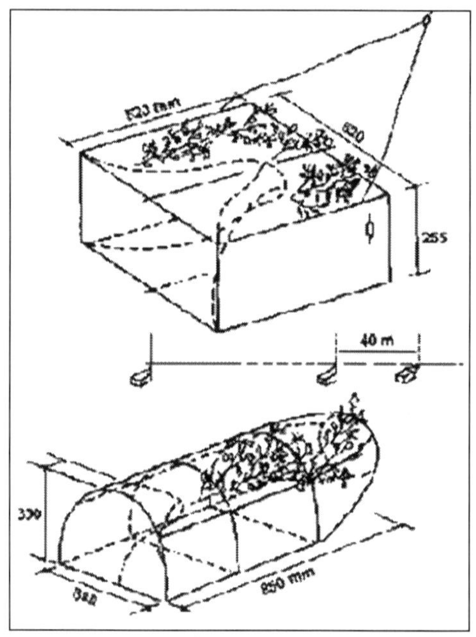

Fig. Cuban fishermen cover their crab traps with mangrove branches to provide a shady enticement. These traps are set in long rows, about 40 m apart.

The use of mangrove boughs to cover trap roofs provides attractive shade and shelter for some species. Cuban cylindrical or box crab traps, made of wire mesh, are lined with boughs and are set out in long rows of 50-60 traps, about 40 m apart. The traps are checked every 2 or 3 days.

The "curiosity trap," used throughout the Caribbean, has curved surfaces, which seem to attract snapper, grouper, jack, and other bottom species. The trap is fished in sandy bottoms near reefs or rocks in water depths between 3 and 80 m and requires no bait. Made of wire netting, it has two funnel entries, which turn down. Since the trap measures approximately 2 × 3 × 1 m and has large funnel openings (11 × 20 cm), fish weighing up to 12 kg can enter. It is usually pulled and cleared every third day.

Variations of Caribbean Island fish traps incorporate exterior funnels leading into a cone-shaped interior funnel that is often turned down to prevent escape. These traps are generally rectangular, although they may be cylindrical or have pointed roofs. Traditionally, cane fibers were used in construction,but marine mesh and galvanized chicken wire are increasingly popular. Fishermen report that such traps catch lobster as well as all types of bottom fish, such as snapper, grouper, and blue runner.

A trash can is used in Hawaii to catch deepwater shrimp. The bottom is cut out of a 10-gallon trash can and is replaced by a 3/4-inch nylon fishnet funnel, which is the only entry. A part of the lid is also cut out and replaced with a 1/2-inch wire mesh. This trap is light, economical, and obtains profitable catches. Whippy bough traps catch fish by the elastic power of a bent bough. When the fish takes the bait, it releases a holding mechanism, which causes the rod to straighten and the catch to be suspended out of the water, clear of predators. This automatic fishing line is common in Java, Thailand, and Europe.

Octopus pots are artificial caves where these animals find refuge, but are free to leave at will. Earthenware pots with handles (Italy, Malta) or without handles (Japan, Korea), are strung singly or on longlines with up to 100 pieces. If the pots are hauled neck down, the octopuses will remain in their shelter. Sections of PVC pipe are used by the Japanese as octopus pots. Two sections of pipe, 15 cm in diameter and 75 cm long, are attached side by side. A 5-cm cement plug bisects each section. This pot offers four dens for octopus and lies flat on the bottom.

An equally effective pot can be made from old tires. The tires arecut in thirds, the rims laced together with wire, and a circular wooden disk nailed to one end. Tire pots have been fished extensively in Venezuela. These pots as well as the PVC pipe pots are fished from bottom longlines at intervals of about 8 m. Buoys mark both ends of the line so that the gear can be easily located and recovered after a fishing period of several days.

Timed-float releases or pop-ups can also be attached to the traps. The pop-ups corrode in saltwater after a given number of days, and the increased rope length permits the previously submerged float to ascend to the surface. This allows concealed traps and reduces poaching, theft, or cut-off floats.

Corrosion is always a serious problem with metal traps. One solution is to use marine mesh, which combines the strength of steel wire with the durability of plastic. The steel or galvanized wire is coated with polyvinyl chloride. Zinc or aluminum anodes can also be attached to metal traps to create a galvanic couple with the iron and double or triple the life of the wire netting. One of the disadvantages of traps is their bulkiness. Some collapsible traps made of wire mesh and plastic are commercially available, but there is a need for local research in this area.

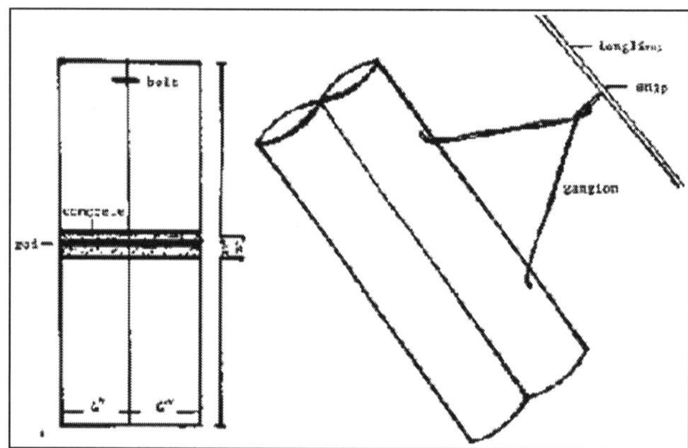

Fig. Sections of plastic pipe can be used as octopus pots. This design creates four chambers, each about 15 × 35 cm, by dividing two pipes in half with a concrete plug.

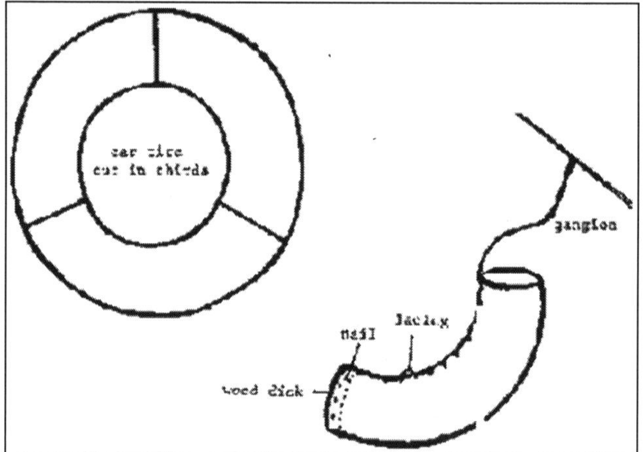

Fig. For use as octopus pots, discarded tires can be cut in thirds, laced into cylinders, and closed at one end with a wooden disc.

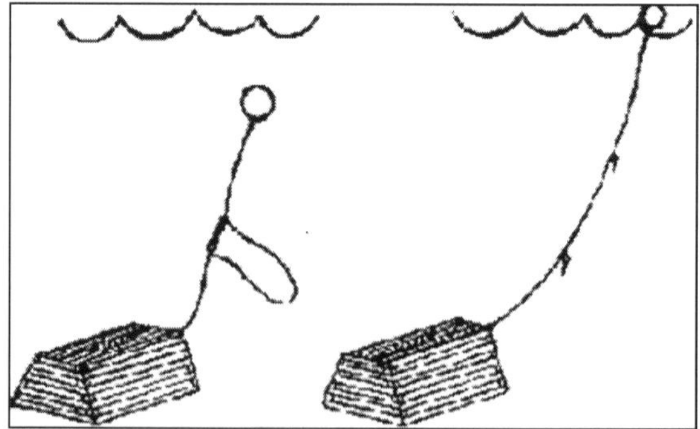

Fig. Seawater corrosion is put to good use with these timed float releases. The thickness of the corrodable band determines how long the marker float stays underwater. For example, traps set with a three-day band on Monday would stay unmarked until Thursday when the float would surface. In the interim, poaching, trap theft, and accidental cutting of the trap line would be reduced.

WEIRS AND TRAP NETS

Large traps can be built in shallow coastal waters to capture sizable schools of fish such as salmon, bonito, tuna, herring, sardines, and cod. Leaders, or wings, guide fish through a slit and into a corral or catching chamber. The small opening acts as a retaining device so the fish do not escape from the corral. Unlike smaller nets, these large structures are not closed at the top. If the walls are constructed of non-textile materials, the trap is called a weir. Pound or trap nets utilize textile netting.

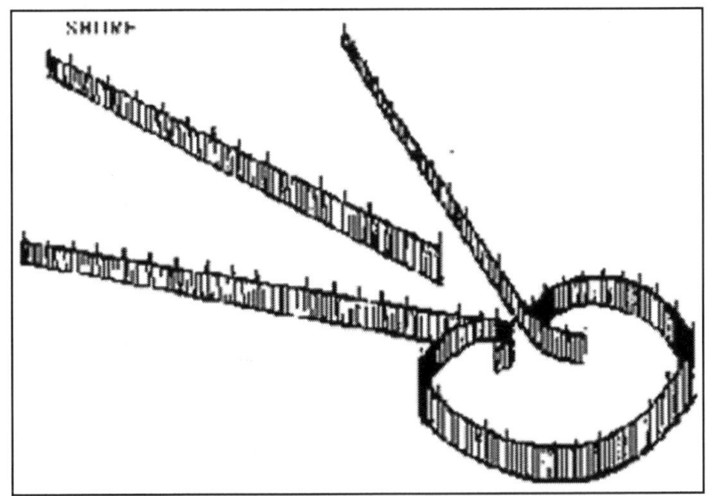

Fig. The bamboo stake trap is set facing the shore. As the tide recedes, shore-feeding fish head for deeper water and the wings of the trap guide them to the center area where they are netted.

Weirs and trap nets have a number of advantages that favor their use in developing countries. They may be constructed of local materials and use little or no energy. Where seasonal runs occur, the catches can be enormous. Indeed, these traps are the only means to intercept large schools without a costly investment in vessels. The fish are live at capture, and therefore in excellent condition. Nevertheless, if the construction material for the traps is netting, costs may be significant. In addition, a great deal of labor is needed to install these devices. These traps are limited to physically suitable areas and to local and seasonal fish behavior. Constant observation is required during the fishing season.

These traps show a tremendous diversity in design. Large bamboo weirs are used in Southeast Asia. The bamboo stake trap consists of long fences of split bamboo, fastened at intervals to large wooden poles driven into the shore bottom. Long wings guide the fish into the heart-shaped center of the trap where they are harvested with nets. It is usually set with the open end toward shore. In trap nets, the netting is usually hung from stakes driven into the bottom, although in deep water anchors and floats may be used. The bottom of the corrals may also be made of netting to facilitate harvesting the fish. In the Caribbean, a simple trap net is used in shallow shelf areas. The length of the wings depends upon the characteristics of the region, but generally the maximum placement depth is 4 m. One of the nylon mesh wings runs from the shore to the corral opening to intercept the migration path of the fish. The other wing is semicircular to prevent their escape and direct them into the corral. From the corral, a narrow opening leads into a smaller holding pool or crib with a net floor to make fish recovery easier. A similar Caribbean trap net is used to capture crabs. The crabs enter the corral through a conical passageway. The lower and upper parts of the corral walls are made of zinc sheeting to prevent the crabs from escaping. This technique is reported to be extremely productive. A Japanese floating trap is another variation on this theme. A floating net cage (25×8×4 m) is anchored over the shallow shelf. One wing extends up to hundreds of meters to the coast and the other to deeper waters. Schools of fish are again directed by the wings through the slits in the cage. Depending on its placement, this floating trap catches demersal as well as pelagic species.

ENTANGLING NETS

Entangling nets are net walls, placed transversely to the path of migrating fish. The bottom of the net is weighted with sinkers while the top is supported by floats. A single-walled net (gill net) is used to gill fish, while a triple walled net (trammel) entangles them.

GILL NETS

A gill net is an upright wall of fiber netting. A fish, of a size for which the net is designed, swimming into the net, can only pass part way through a single mesh.

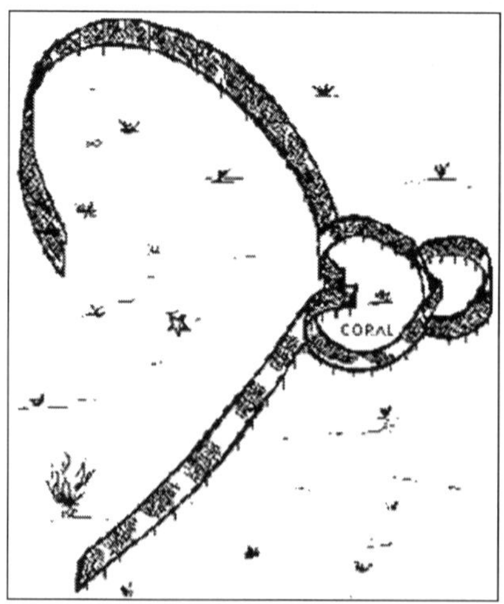

Fig. This Caribbean trap net is set to capture fish swimming parallel to the shore. One wing of netting extends from the shore to the corral and the second is placed in a semicircle to deflect escaping fish.

As the fish struggles to free itself, the net twine slips in back of the gill. The fish is thus gilled and can go neither forward nor backward. Various mesh sizes are employed, depending on the species and size of the fish to be caught.

TRAMMEL NETS

Trammel nets have three panels of netting suspended from a common row of floats and attached to a single bottom line. The two outside walls of netting have a mesh larger than the targeted fish, and the interior netting has a smaller mesh size. The inside net hangs loosely between the two outer nets. A fish striking from either side passes through the large mesh outer panel, strikes the smaller mesh interior panel, and carries it through the opposite large mesh panel, forming a sack or pocket in which the fish is trapped. A trammel net is often fished by drifting.

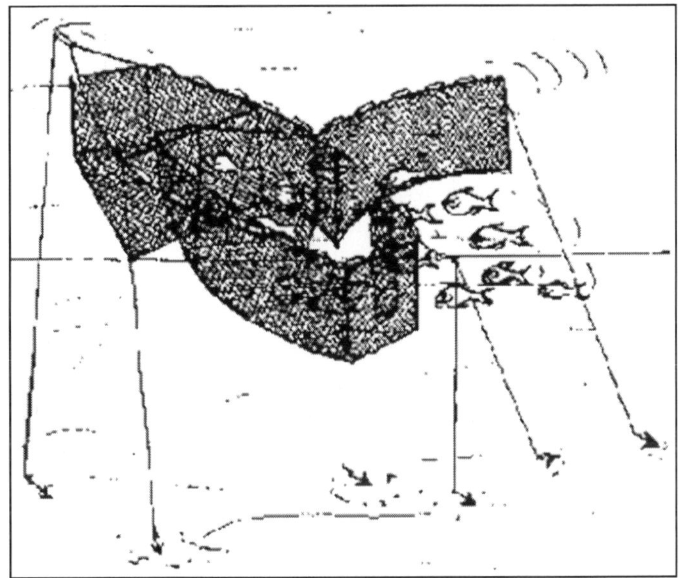

Fig. A Japanese floating trap is anchored in shallow water. Schools of fish are intercepted and diverted into a net cage.

These nets may be used at the surface, in midwater, or at the bottom. Although a single net may be deployed, usually a row of nets are set. A fisherman may choose to anchor his net or allow it to drift. In intertidal areas, the nets may be driven into the bottom and the fish collected at low tide. Taiwanese fishermen tie scare ropes around their bodies and swim as a group toward a fixed net. Fish in the swimmers' path are frightened into the net.

To be most effective, a net should be invisible to the fish. In the past, cotton nets were dyed different colors to match the background. Nowadays, transparent monofilament nets are mainly used for entangling nets, though more visible but softer continuous filament nylon nets are still popular in some fisheries.

Monofilament fibers are less elastic and stiffer than continuous multifilament nylon fibers. Thus, although the former are more efficient in catching the fish, the latter hold them better. The great advantage of entangling nets is their selectivity. The way the net is hung and its depth determine the species of fish captured. The shape and size of the mesh also select fish species of a specific size. Fish whose girth is smaller than that of the mesh opening are able to swim through.

Even a lone fisherman can manually deploy a gill net or a trammel net from a small craft. Floats and sinkers can be made of local materials such as bamboo, bottles, cement, or stones, although manufactured equipment might be more efficient. These advantages are balanced by serious liabilities. Entangling nets cost more than hooks or traps. They require a high degree of maintenance, and picking the fish out of the net is labor intensive. Since the fish are usually dead when harvested, they will be of lower quality.

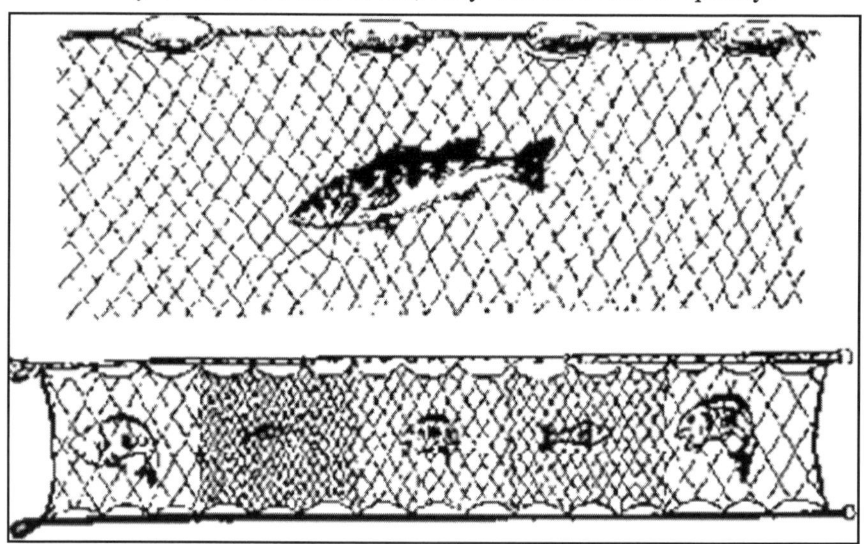

Fig. The mesh size of a gill net determines the size of the fish caught.

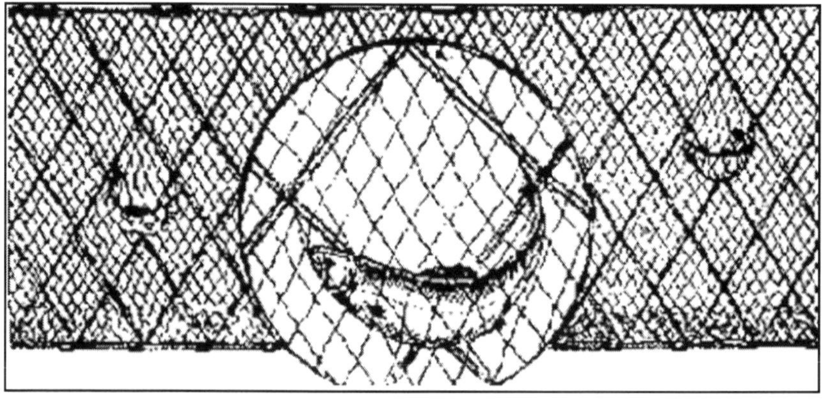

Fig. Trammel nets capture fish in a pocket of netting.

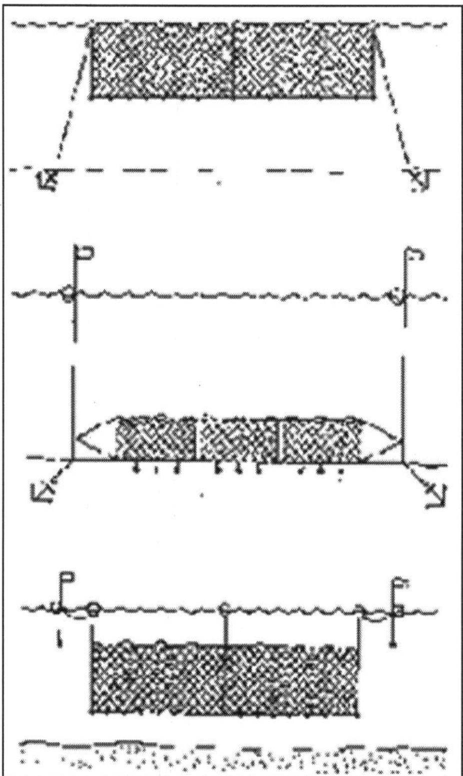

Fig. Entangling nets can be used at the surface, midwater, or on the bottom, either anchored or free-floating, depending on the target fish.

Ghost fishing with nets made of synthetic fibers is a problem. If lost, the nets continue to trap fish, because the fibers are not biodegradable. To avoid this, the twine holding the netting to the floats should be made of natural fibers, which will rot in time.

STATIONARY LIFTNETS

Liftnets are lifted from the water at the moment when the sought-after fish have gathered over them. These nets can be installed on boats or on the shores of rivers, lakes, lagoons, and estuaries. Lifting power may be provided by pulleys or weighted levers. In South India, liftnets are operated off the beaches and lifted with counterweights. At night, lamps are hung from the crossbars to attract fish.

Light is also used to attract fish to liftnets operated from shipboard. In the Caspian Sea, Soviet fishermen use small circular liftnets equipped with underwater electric lamps to catch anchovy. A larger variation of liftnet is called a blanket net. Operated from the ship's side, it can almost be as wide as the vessel's length. Four-boat liftnets are common in Scandinavian and Southeast Asian fisheries. The net is lowered or raised by all the boats in unison.

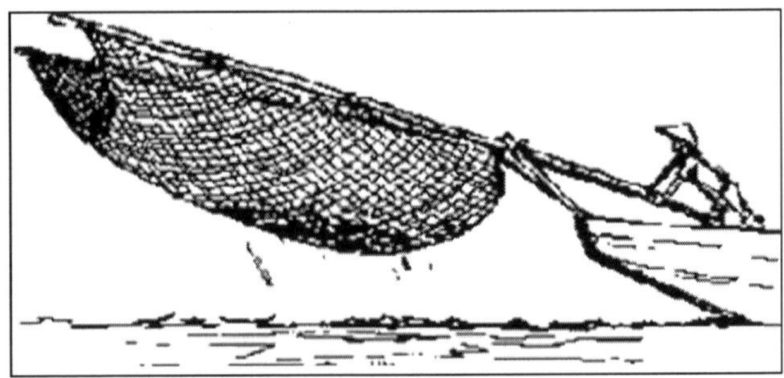

Fig. Liftnets can be used from small boats or from= the shore. Fishermen bait the net area and wait for fish to gather. At night, torches or lamps can be used to attract fish.

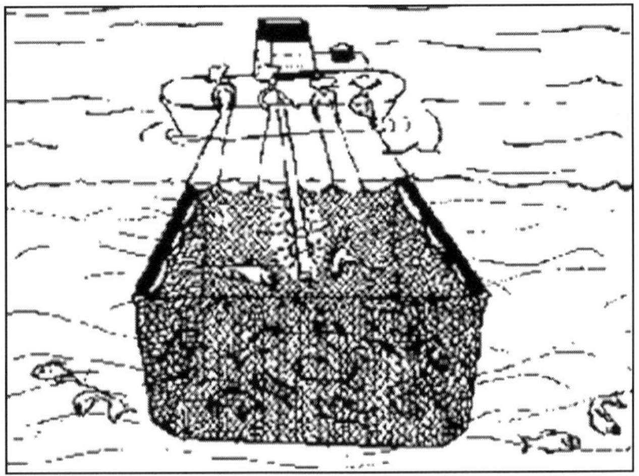

Fig. On larger boats, electric lights are used with liftnets to capture anchovies and sardines.

The Lake Tanganyika liftnet is suspended from hauling ropes at the four ends of a catamaran 3.35-6.0 m long. The net is shaped like an inverted pyramid and has a stretched depth of 12 m. Kerosene lamps are used to attract the fish over the net opening that is suspended between the two hulls. After the fish have been concentrated, the net is quickly hauled through the open area between the hulls.

Liftnets can be cost effective and labor effective when set under specific conditions (narrow passages and rivers) or when attracting fish with chum or light. The catch is alive and, therefore, of good quality. This fishing technique, especially in combination with light attraction, could be used in many areas, either for consumable fish or live bait.

ACTIVE GEAR

Active gear has to be moved, dragged, or towed in order to capture fish. This usually requires engine-propelled boats and usually involves additional investment over passive or stationary gear.

TROLLING

Trolling lines are simple hooked lines that are trailed from a moving vessel at a controlled depth. Bait may be artificial or natural and attracts predator fish that see what appears to be a smaller fish thrashing and turning in the water. The lure may be nothing more than a colorful piece of cloth, a small bunch of feathers, or a piece of skin from the bait fish, but it must be carefully adapted to local conditions and fish species and size. The use of outriggers can increase the number of lines that can be trolled and helps keep them from becoming tangled.

Trolling offers numerous advantages to small-scale fisheries. Multiple trolling can be performed from a reasonably small craft. Changing sinker weights allows fishing at graduated depths. Lures can be made of local materials and easily changed for the target species. The use of artificial bait avoids the capture or purchase of live bait.

Trolling is also an excellent auxiliary method and can be used as the vessel is going to or returning from other fishing grounds. Eligible areas include inshore or offshore waters, and target species may be pelagic or demersal. The introduction of trolling does not require high skills or a large investment in gear.

Little labor is necessary in this fishing art. Sails are an ideal form of propulsion for this fishing method.

Fig. Additional fish can be caught by trolling between the fishing grounds and landing. The fishing depth can be adjusted by changing sinker weights.

A number of innovations can make trolling more efficient and save labor. Simple hand and electric reels can make work easier and allow more fish to be caught in a given time.

Large and small umbrella rigs permit the fishermen to have several hooks on a trolling line. For many species umbrella rigs are more attractive than conventional lures, perhaps because the multiple lures create the illusion of a school of fish. The number of lines a vessel can troll can also be increased by towing two smaller boats to spread the lines over a larger area.

Depressors are used by Japanese fishermen for midwater trolling. These are flat, small boards weighted with lead at the front edge. The depressor itself is towed by a line attached near its front edge. Besides submerging the line, the board also wobbles, so that the hook jumps or jigs. When a fish takes the hook, the board tilts and rises to the surface.

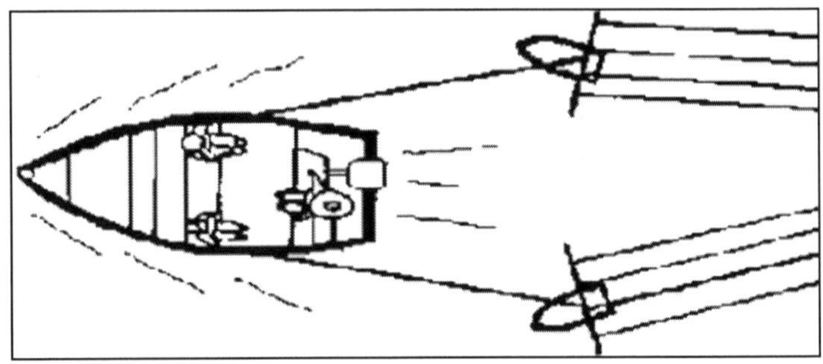

Fig. Multiple trolling lines can be used if poles, floats, small boats are used to spread the lines.

Another trolling variation, the fish kite, is popular in Micronesia, Melanesia, and Indonesia. The kite is flown up to 100 m behind the boat, and its tail line carries a ball of cotton or a piece of sharkskin as bait. The kite is maneuvered so that the bait continuously bobs on the surface of the water and induces garfish to snap at it.

JIGGING

The jigging technique involves mainly catching fish by impaling them with special hooks. In jigging, the line must be jerked to pierce the fish. Generally, the sharp hooks are weighted so that when they are pulled up, there will be sufficient momentum to penetrate the fish.

In some cases, regular baited hooks are jigged manually or mechanically to attract attention to the bait. Special reels can be used to impart a jigging action to the line.

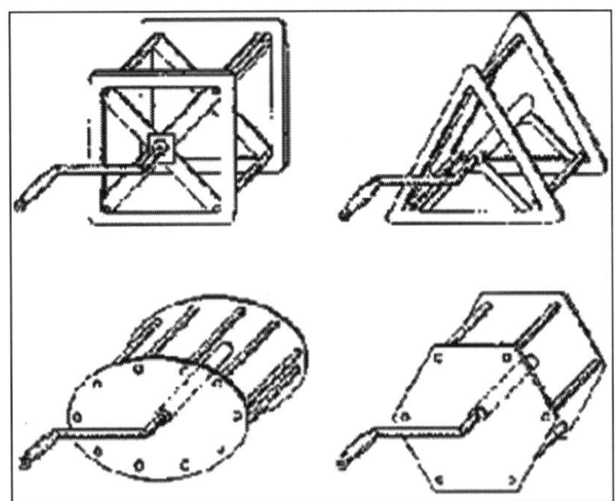

Fig. Specially designed reels can give a jerking action to lines as they are hauled; they improve results in jigging.

Rippers or jigs are especially used to catch slow-moving fish that are spawning. They are also widely used throughout marine fisheries. The Norwegian Juksa-line catches cod by jigging, and the Turks jig for bluefish in the Bosporus. Special hooks are used in east Asia and the Mediterranean to jig squid. The potential exists for significant expansion of squid jigging by light attraction.

Jigging is a low-cost, low-energy technique that does not require bait. The live catch is easily brought into the boat. At the same time, it is labor intensive and time consuming, unless relatively expensive jigging machines are employed. Jigging requires knowledge of the local area to determine where and when it can be used.

SEINING

Seines are long nets with meshes small enough to prevent the desired fish from gilling (filtering nets). They are generally set in a semicircle and dragged over a smooth bottom by means of long ropes (sweeps). In this way, the fish are herded into the net and hauled onto the beach or on board.

BEACH SEINES

Beach seines are especially appropriate for catching seasonal pelagic species as they feed near shore. They are most often set from the boat. One end remains on shore, while the rest of the net is set in a curved path and brought back to the beach.

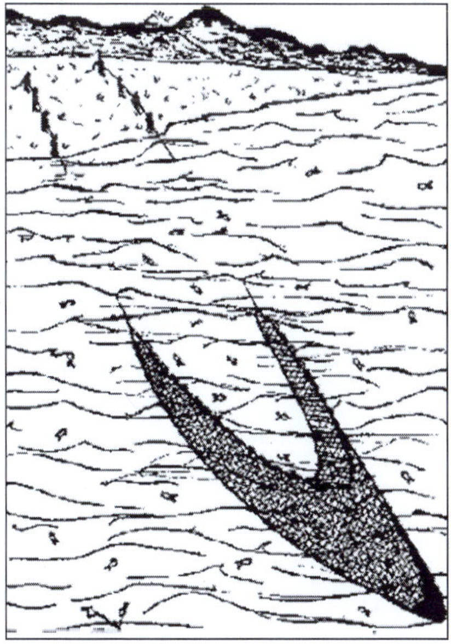

Fig. Using large beach seines is a group effort. One end is held on the beach while the other is rowed in a curve back to the beach and hauled in.

Once the second drag line is delivered to shore, the hauling begins. The bottom and water surface act as natural barriers for the fish encircled in the net. The wings may often be hundreds of meters long. Large beach seines, however, are costly, and their use is restricted to large stretches of smooth, shallow bottoms with fairly mild surf. The net is species-indiscriminate and may catch juveniles of large-sized fish. Small two-man beach seines are often used for catching live bait or small fish. These tend to have a uniform small mesh.

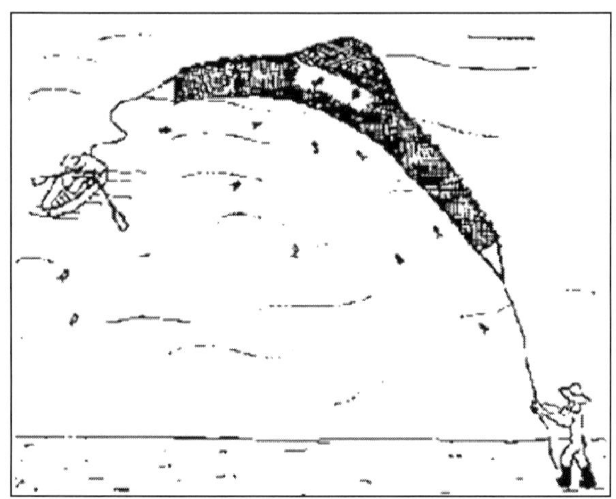

Fig. Smaller beach seines can be a two-person operation, useful for catching bait fish.

Beach seines have the potential for increased motorization and mechanization. Shore-anchored pulleys, tractors, jeeps, or even animals could be used to make hauling easier.

BOAT SEINES

Boat seines are set and hauled from a boat. A vessel anchors one end of the seine and sails in a circle, releasing the net, and returns to the anchor. The net is then hauled into the boat.

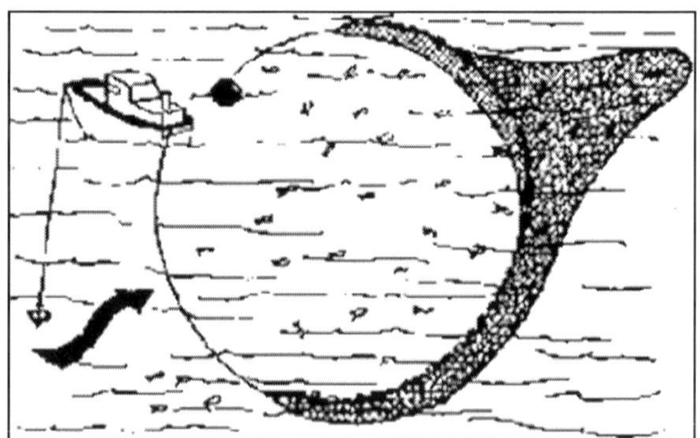

Fig. Large boat seines are used by anchoring one end of the seine, sailing in a circle back to the anchor point, and hauling the net.

With small seines, this fishing technique can be used by smaller vessels without mechanization almost anywhere there is a smooth bottom. The high skill involved in net design and the cost of its construction are liabilities.

Fig. Small boat seines can be handled by two fishermen without mechanization.

Boat seines have the potential of bringing small-scale fishermen to previously unexploited resources. However, any introduction of modern technologies (motorization of the boat, rope and net haulers, storage of nets and cables on reels) that would enlarge the area covered and could increase the catch, involves capital investment and an increase in running costs.

PURSE SEINES AND RING NETS

Purse seines are characterized by a line at the bottom of the net that is used to close off this escape route. The purse seine can be set with one or two boats and must be fished quickly. Those that are operated with two boats are called ring nets. Light may also be used to attract the target species. Purse seines are highly mobile and can capture whole large schools of pelagic species that gill nets and beach seines could not. Hauling can be done manually, and the catch is live. Nevertheless, purse seines are costly and require highly skilled operators. Purse seining with two boats (ring netting) enables small, artisanal fishing craft to take advantage of this method.

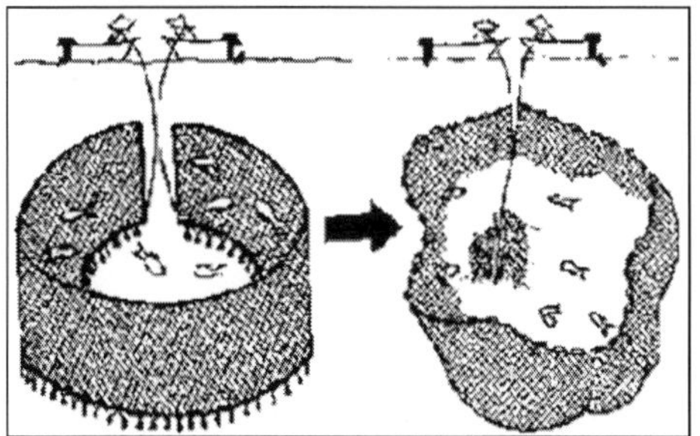

Fig. Purse seining with two boats (ring netting) allows smaller boats to use this technique.

BOTTOM TRAWLING

Trawls may be towed behind one or two boats or, in shallow waters, even dragged by a fisherman. Trawl nets generally have a cone-shaped body with a wide opening between two wings. In bottom trawling, the net is towed on the bottom in order to capture shrimp and demersal fish.

Fig. Trawl nets can be pulled by one or two boats or, in shallow water, dragged by a fisherman.

PAIR TRAWLING

Pair trawling uses two small boats to tow the trawl, one on each side. Having two boats keeps the trawl net open. This method also permits boats with small (5 hp) engines to trawl and allows small-scale fishermen to compete with larger trawlers.

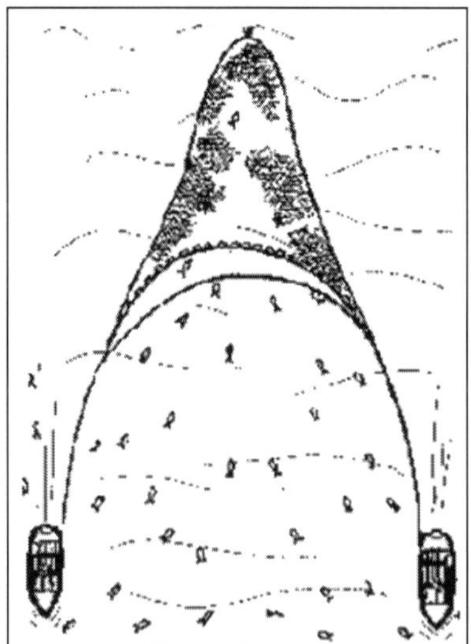

Fig. Boats without enough power to trawl singly can often trawl in pairs. Using two boats allows a wider area to be covered and makes it easier to keep the net open.

With the same total horsepower, more fish can be caught with pair trawling than if a single boat tows the net. Whereas the noise from a single engine directly in front of the trawl net can frighten fish from the path of the net, the noise from two engines on either side of the opening will scare some fish towards the center, directly into the net.

Pair trawling has limitations. Two boats must cooperate and work as a team. The fishing area is limited to smooth bottoms. Even in ideal areas, the net can be damaged or lost on a wreck or a rock. The value of the catch must be at least equal to the sum of the value of the two vessels' catches if they fished alone.

E the boats have engines stronger than 8 hp. they are strong enough to tow sweeplines.These lines are made of heavy rope and are towed on the bottom in front of the wings of the trawl net. They serve to scare fish from a wider area into the net.

SINGLE BOAT TRAWLING

A single vessel with an adequate power source may also tow a trawl, but otter boards or a beam are required to open the net horizontally. Beam trawls are the simplest trawls and are used primarily to capture flatfish and shrimp. The horizontal opening for these nets is provided by a beam made of wood or metal that can measure up to 10 m in length.

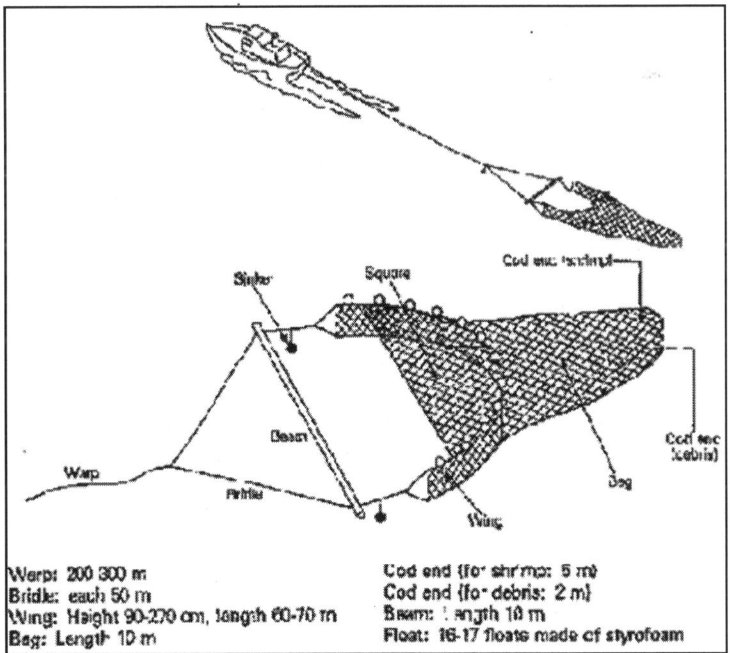

Fig. Beam trawling is accomplished from a single boat. An 8-to 10-m pole (beam) is used to keep the net open horizontally to capture flatfish or shrimp.

Smaller beams, about 2 m in length, are used with rowboats in Portuguese rivers. Although small beam trawls might be used by artisanal fishermen, they obviously lack the fishing spread of larger trawls, which require power and mechanization.

Otter trawling is a more complex fishing system. These trawling nets have their horizontal opening maintained by the shearing action of the heavy otter boards. Demersal or pelagic species can be captured by this fishing method in shallow waters.

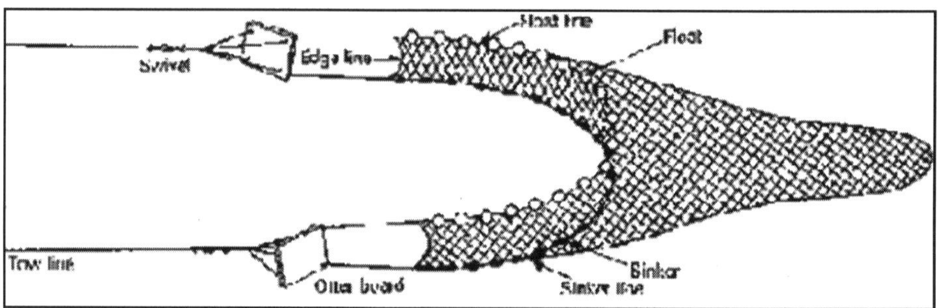

Fig. In otter trawling, two flat (otter) boards are used at either end of the net to hold it open.

Otter trawling gives fishermen broad access to marine resources. But the high costs, large energy requirements, and the specialized skills required to maintain the equipment and use it effectively make it feasible for small-scale fisheries only under very favorable conditions.

The minimum power for an otter trawling boat is 30-40 hp with a relatively high gear ratio (low propeller rpm) and a large propeller diameter to provide maximum towing power.

ELECTRONIC EQUIPMENT

Much marine electronic equipment was initially developed for military use in communications, navigation, and underwater reconnaissance during World War II.

Postwar growth in the electronics industry resulted in lower costs for this type of equipment and ocean-going fishermen began to use it. As costs decreased even more, the market has broadened to include smaller-scale commercial and sport fishermen. Although probably still beyond the reach of most individual fishermen in developing countries, some of this equipment may be cost effective for shared use in villages or cooperatives.

Perhaps the most useful for nearshore fishermen would be aids to fish location. The simplest of these is an electronic thermometer. Seawater temperature can markedly affect fish-feeding habits, and in thermally stratified water, species may concentrate at depths based on temperature. In addition to the value of knowing absolute temperature and its relationship to fish feeding and depth, changes in temperature are also important. Seawater temperature can remain constant over a wide area; a change of a degree or even less can indicate an upwelling or current boundary where fish may cluster. Stem thermometers that rely on liquid or metal expansion and contraction for temperature readings are not responsive enough for this application. Simple digital readout electronic thermometers can display instantaneous temperature changes of tenths of a degree. These are available for less than $100.

Another valuable device is an electronic depth recorder. These can indicate water depth, bottom formations, and fish locations. Boats need travel no farther than is necessary to detect fish. Nets and lines can be set and hauled with greater efficiency.

Rocky bottoms potentially damaging to trawls can be detected. The results of a properly used depth recorder can be dramatic and should have a direct and visible economic benefit. To use this equipment, a fisherman must install a transducer on the hull. A method of installing this unit on temporary brackets has been developed to allow its ready transfer from vessel to vessel. Costs for these echo sounders range from $200 to $600.

Although excellent Loran and satellite electronic navigation aids are available, their costs are prohibitive. Where appropriate radio stations operate, inexpensive radio direction finders can be used to plot positions and plan courses.

ACTIVE GEARS

Active fishing methods are especially suitable for sampling large proportions of the whole fish stock or large numbers of fish. The term “active” means that the fishing gear is dragged through the water by human, animal or engine power. In most cases the efficiency of active gear is considerably higher than that of “passive” gear, such as gillnets and traps, which rely for their efficiency on the movement of the fish themselves.

BEAM TRAWL

In this type of trawl, the mouth or opening of the net is kept open by a beam, which is mounted at each end on guides or skids, which travel along the seabed.

The trawls are adapted and made more effective by attaching tickler chains (for sand or mud) or heavy chain matting (for rough, rocky ground) depending on the type of ground being fished.These drags along the seabed in front of the net, disturbing the fish in the path of the trawl, cause them to rise from the seabed into the oncoming net. Modern beam trawls range in size from 4 to 12 m (weighing up to 7.5 tonnes in air) beam length, depending on the size and power of the operating vessel.

DEMERSAL OTTER TRAWL

BOTTOM TRAWL

The demersal or bottom trawl is a large, usually cone-shaped net, which is towed across the seabed. The forward part of the net – the 'wings' – is kept open laterally by otter boards or doors. Fish are herded between the boards and along the spreader wires or sweeps, into the mouth of the trawl where they swim until exhausted.

They then drift back through the funnel of the net, along the extension or lengthening piece and into the cod-end, where they are retained.The selectivity of trawl fisheries may be increased by the use of devices known as separator trawls. Separator trawls exploit behavioural differences between fish species and can be used, for example, to segregate cod and plaice into the lower compartment of the net, whilst haddock are taken in the upper part. The mesh size for the two compartments can be altered according to the size of the adult fish being targeted. Insertion of square mesh panels also improves selectivity of the net because square meshes, unlike the traditional diamond shape meshes, do not close when the net is towed. Sorting grids are compulsorily fitted in nets in some prawn and shrimp fisheries to reduce bycatch of unwanted or non-target species, including small prawns and shrimp.

Depending on the depth of water fished and the way in which the gear is constructed and rigged, trawling may be used to catch different species. Trawls can be towed by one vessel using otter boards, as in bottom-trawling, or by two vessels, each towing one warp, as in pair-trawling or more than one trawl can be towed simultaneously as in multi-rig trawling.

PELAGIC TRAWL

When trawling takes place in the water column or in mid-water between the seabed and the surface, it is referred to as mid-water or pelagic trawling. Pelagic trawls target fish swimming, usually in shoals, in the water column i.e. pelagic species.

These include seabass, mackerel, Alaska pollack, redfish, herring and pilchards for example. Their effectiveness relies on traversing a considerable volume of water, and consequently nets are larger than bottom trawls and require a large vertical and horizontal mouth opening to provide net stability and capture large shoals of fish. The length of time the net is towed through the water is shorter than in bottom trawling in order to capture the shoals of fish the net passes through. To handle the large amounts of fish, pumps are used to transfer the catch from the cod-end to the boat.In mid-water pair trawling the otter boards are replaced, and the mouth of the net kept open, by a pair of trawlers. This enables vast nets, often ¼ mile wide and ½ mile long, to be towed through the water column to capture the fish.

SEINE NETTING

This is a bottom fishing method and is of particular importance in the harvesting of demersal or ground fish including cod, haddock and hake and flat-fish species such as plaice and flounder. The fish are surrounded by warps (rope) laid out on the seabed with a trawl shaped net at mid-length. As the warps are hauled in, the fish are herded into the path of the net and caught.Effectiveness is increased on soft sediment by the sand or mud cloud resulting from the warps' movement across the seabed.

This method of fishing is less fuel-intensive than trawling and produces a high quality catch, as the fish are not bumped along the bottom as with trawling.

PURSE SEINING

This is the general name given to the method of encircling a school of fish with a large wall of net. The net is then drawn together underneath the fish (pursed) so that they are completely surrounded.It is one of the most aggressive methods of fishing and aims to capture large, dense shoals of mobile fish such as tuna, mackerel and herring.

TROLLING

Trolling involves towing baited hooks or lures through the water. The method is particularly suited to the capture of pelagic species of high individual value. Examples include tuna (albacore and skipjack), barracuda and salmon.

HARPOON

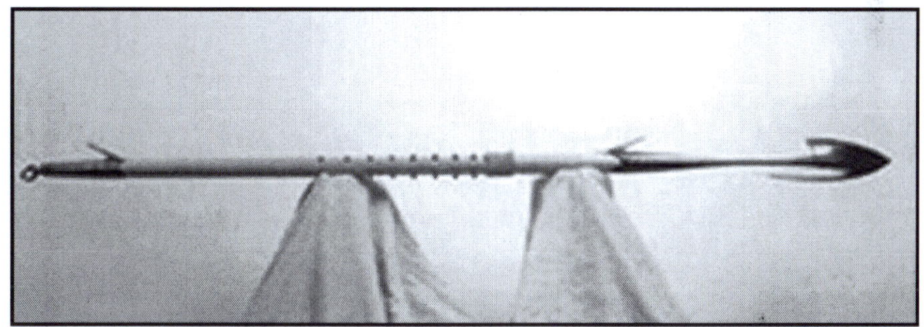

This method is used for fish having high individual value such as swordfish and bluefin tuna. Harpooning is a completely selective fishery, since the target must be seen before striking, so the size and hence age can be determined and only mature fish taken.

LIMITATIONS

The introduction of fishing gear and methods to an area, whether these methods are technically new or simply new to a given region, is not without dangers. Gear and methods are highly site specific. The fishing arts that have developed in a region are usually the best suited for the species and size desired, the given marine conditions, and the community's economy and structure. New methods may often be inappropriate and rejected by local fishermen.

Introduced gear may be too costly for the local economy to sustain. What appears to be economical to an outside observer is often impossibly expensive for the fishing community unless increased credit is made available. Innovations should be tested on a pilot scale to ensure that they are economically viable. Many coastal waters are overfished. Upgrading the gear and making it more efficient increases the risk of depleting the fishing stocks even more. Therefore, the introduction of any new gear or methods must be accompanied by proper monitoring and protection of the marine resources.

More sophisticated fishing arts may require training for the fishermen. At the same time, modifications in gear may necessitate simultaneous improvements in the design, power, and size of fishing vessels. Almost all marine electronic equipment requires 12 volts DC to operate, although current requirements are fairly low, less than 10 amps. Because most of this equipment is not designed for user repair, maintenance is a more severe problem. The most practical approach is to minimize the number of different models in use.

RESEARCH NEEDS

A persistent concern when introducing new gear or methods is that of exceeding sustainable fishing yields, thereby making the fisherman's lot temporarily better but ultimately worse. Research on simple methods for determining fish populations and their regenerative capacity could be very valuable. New gear or methods could then be used without fear of overfishing.

In situations where specific gear is potentially valuable but too costly for a fishing community, adaptive research on local manufacture using local materials could be an alternative. More information on combinations of attraction and capture methods is needed. The use of fish-aggregating devices or light in conjunction with complementary traps or nets could improve the catch of specific sizes species.

FISHING GEAR AND FISHERY ACTIVITY

There are many different types of fishing gear. Some gear has been adapted to certain species on the basis of the species' special characteristics such as their behaviour, their feeding and spawning migration patterns. Changes in fishery activities throughout the year are due to biological and climatic conditions. Active fishing methods have been employed ever since the stone age and have developed over the ages to give us the wide variety of fishing gear we have today.

For Norwegian waters, the fishing gear can be divided into the following main groups:

1. Gillnet (bottom-set gillnet, midwater gillnet and drift net)
2. Hook-and-line gear (lines, jigging and trolling lines)
3. Trawl (bottom, pelagic and semi-pelagic trawl)
4. Seines (Danish·and Scottish seines)
5. Closing nets (Purse seines, shore seines)
6. Other fishing gear (Such as lobster/crab pots, fish traps and shell scrapers)

Within each main group of fishing gear, such as nets and trawls, there are many different types and sizes. It would not be relevant to provide a detailed description of these in this summary. Thus, in the following, the fishing gear has been divided into two main groups and described in sufficient detail to provide a summary of the most important fishing gear that one can expect to come across in the geographical areas where seismic surveys are conducted. The summary also describes fishing gear used by foreign vessels in the Norwegian Economic Zone. The main groups of fishing gear are divided into passive and active gear.

Def. passive fishing gear: A fishing device consisting of gear where the fish must seek out the gear in order to be caught, such as longlines and gillnets.

Def. active fishing gear: A fishing device consisting of gear where the gear must seek out the fish to catch it, such as trawls, closing nets and seines.

The most common species in Norwegian waters can be divided into two main groups:

Pelagic fish: Herring, mackerel, blue whiting and capelin.

Demersal fish (and shellfish): Cod, haddock, saithe, prawn, redfish, Greenland halibut, Tusk, ling, blue whiting, greater argentine, spotted catfish, sandeel, Norway pout, plaice, whiting, crayfish, monkfish and wolfish.

Pelagic fish generally live in the open ocean, though they may be found near the seabed during certain periods. Demersal fish mainly live on or near the bottom of the sea, but may also appear in the open sea on occasion, shellfish excepted.

GILLNET

Gillnet fishing has long traditions. Today it constitutes the most important fishing device for much of the Norwegian fishing fleet. Gillnets are principally used by the coastal fishing fleet, but also by larger ocean-going vessels. Fish commonly caught in gillnets include cod, saithe, Greenland halibut, redfish, ling and monkfish. The most common fishing operations that come into contact with seismic surveys are Greenland halibut, cod, saithe and monkfish fishing. Gillnet fishing takes place in various degrees from south to north in the Norwegian Economic Zone. A gillnet is roughly speaking a piece of net (mesh) with floaters attached to the top and weights at the bottom. The length and height of the gillnet vary according to the type of fish one wants to catch, as does the mesh size. Furthermore, a distinction is made between various types of gillnet fishing such as bottom-set gillnet fishing, midwater gillnet fishing and drift-net fishing. The latter two are not considered relevant in this connection as such gillnet fishing rarely takes place in areas with seismic activity, and have thus not been included in this description.

BOTTOM-SET GILLNET

When fishing with bottom-set gillnets a number of nets are tied together to form what is referred to as a chain of gillnets. A chain of nets usually consists of between 10 and 40 gillnets, but may also be longer. One gillnet is usually around 28 metres long, though the length may vary according to the type of gillnet.

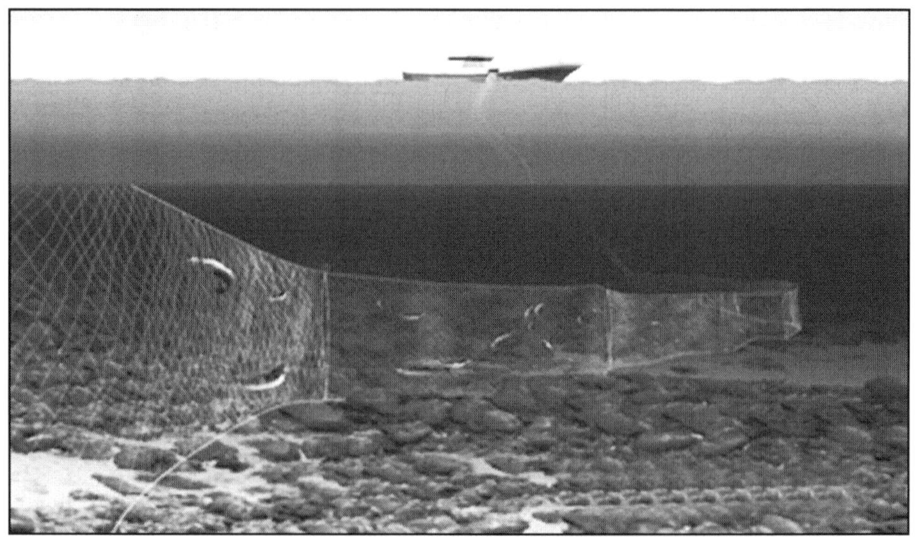

Fig. Illustration of a chain of gillnets.

The chain of gillnets is anchored to the seabed when set. Usually stones, iron grapnels or anchors are used as anchorage, depending on the anchoring required in the specific area to prevent the chain of gillnets from drifting with the current.

The weight of the anchorage varies, but usually ranges from 20 to 120 kilos, depending on the depth and currents. A vertical line is extended from the surface down to the anchorage, often referred to as a dropline (terms and expressions may vary according to region and regional dialect). The length and thickness of the line are adapted to the size of the vessel and to the depths and currents. Sometimes lines are used that are about 1.75 times longer than the depth.

This means that surface buoys can be observed outside the given position of the chain of nets. It is very important to take this into account to avoid collisions with the fishing gear. The dropline is often marked by one or several corklines, (buoy in front of the main float), floats and a pole with a flag and/or reflectors.

The flags indicate which end of the chain that is seen In periods of strong currents, the surface buoys may be pulled under the surface and re-emerge when the currents subside.

At greater depths and "hill crests" the chain of gillnets is usually set with a so-called "loose end". This means that only one end of the chain is anchored.

Coastal gillnet fishing vessel (smack)

Ocean-going gillnet fishing vessel(combined longline and gillnet)

Fig. Illustration of net fishing vessels.

As described above, the length of the chain of gillnets varies according to the type of fishing (main species of fish). However, it is rare to see just one chain in the same area.

More often than not, one vessel will have several chains of gillnets in the same area and during intense fishing for species such as cod and saithe, many vessels with multiple chains will fish in more concentrated areas. Below follows an illustration of an activity pattern for bottom set gillnet fishing (chain of nets) in the "Egga Edge" area (continental slope).

HOOK-AND-LINE GEAR

Hook-and-line gear is a term used for various fishing gear which can, generally speaking, be divided into the main groups "longlines" and "trolling lines". Within the category longlines are a number of lines such as mid-water lines, stake lines, boulters, bottom-set longlines, etc. Only bottom-set longlines are relevant to this description. As regards longline fishing, we have experienced that there have been some conflicts between the fisheries and the seismic activities regarding the use of area. We will therefore describe longline fishing in detail to improve the understanding of how this type of fishing is conducted. Within the group "trolling lines" we find it relevant to describe mackerel jigging.

BOTTOM-SET LONGLINES

Fishing with bottom-set longlines has long traditions in the Norwegian fishing fleet and is an important fishing gear for both the coastal fleet and the ocean-going fleet. Some vessels using lines have also been designed for gillnet fishing in certain periods. The main species for longline fishing comprise cod, haddock, Greenland halibut, ling, tusk and catfish. However, the coastal fleet's line fishing for cod, haddock, ling and tusk has a smaller contact area with areas where seismic surveys are taking part.

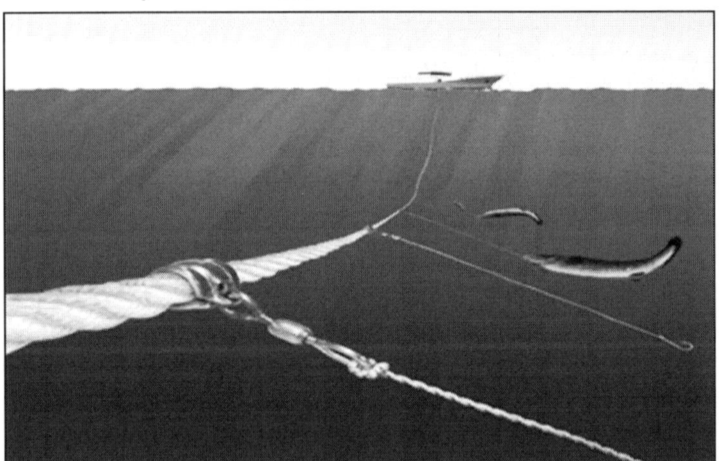

Fig. Illustration of longline during hauling.

As its name indicates, bottom-set longlines are set on the seabed. A line is roughly speaking a length of rope (mainline) onto which hooks are attached at regular intervals along the rope. The rope or line attaching the hook to the main line is called a snood. The size of the hooks differs mainly between hooks used for coastal and ocean fishing. A row of lines or a stub which it is sometimes also called, is set up in the same way as a chain of gillnets. The stub is divided into a given number of branch lines and one line has about 200 hooks. The length of the stub varies according to fish concentrations and seabed conditions.

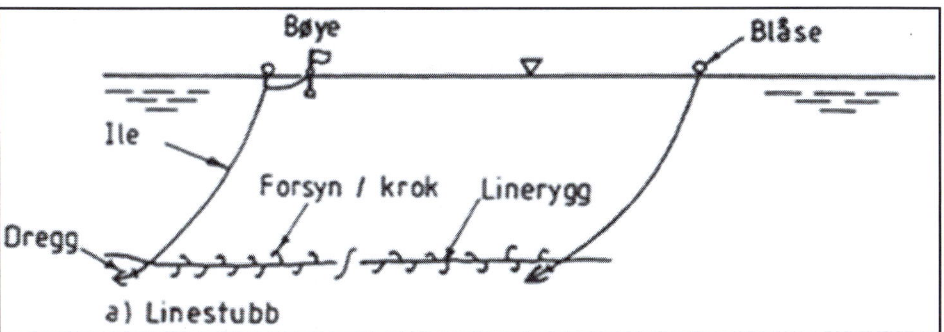

Fig. Illustration of a row of lines

The row of lines is anchored to the seabed using a normal iron grapnel or an anchor, depending on the anchorage required in the specific area to prevent loose ends from drifting with the currents. The weight of the anchorage varies but ranges from 20 to 80 kilos, depending on the depth and currents.

A vertical line is extended from the surface down to the anchorage, often referred to as a dropline (terms and expressions may vary according to region and regional dialect). The length and thickness of the line have been adapted to the size of the vessel and to the depths and currents. Sometimes lines are used that are about 1.75 times the depth. The dropline will always be longer than the depth. This entails that surface buoys can be observed outside the given position of the chain of nets. It is very important to take this into account to avoid collisions with the fishing gear.

The dropline is often marked by one or several corklines, (buoy in front of the main float), floats and a pole with a flag and/or reflectors. In periods of strong currents, the surface buoys may be pulled under the surface and re-emerge when the currents subside. At greater depths and "hill crests" the chain of lines is usually set with a so-called "loose end". This means that only one end of the chain is anchored.

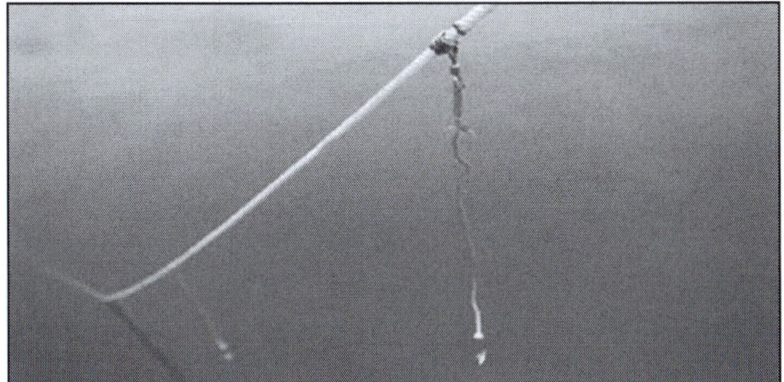

Fig. Photo of a main line.

The longline fishing itself takes place in basically the same way as gillnet fishing. However, there are bigger differences between coastal vessels and ocean-going autoline vessels in the way they fish. The fishing gear extends farther than chains of nets. On an average 4-6 week's trip, an ocean-going autoline vessel is usually in operation 24 hours a day, seven days a week. An autoline vessel sets and hauls between 30,000 and 50,000 hooks per day. It is not usual to set all the hooks on one stub (35 – 40 nautical miles). The stubs are usually divided into lengths of 8 – 10 nautical miles. Some fisheries divide the stubs into even shorter rows, increasing the number of stubs in the area.

A coastal vessel may also operate on a 24-hour basis using an autoline (automatic baiting), but their trips are shorter (from one to a few days) due to the need to deliver fresh fish. However, most of the coastal line fishing is based on manual baiting on shore. The vessel will bring ready-baited lines to the fishing grounds and set the line in one or several stubs. After leaving the lines in the water for some time (variable), the line is hauled in and the vessel returns to shore. The time the vessel takes from departure to arrival is referred to as one "sjøvær" (one fishing day).

As part of its requisite operation pattern, an ocean-going autoline vessel will always keep moving into new areas. If not, it will soon experience a reduction in catches. This means that rows of line are set in a "new and unused" area, often in parallel to where the previous row was hauled in. Whether the vessel casts the line again immediately after it has been reeled in or after hauling in several lines will depend on the individual vessel and fishery. An ocean-going autoline vessel keeps moving into new areas as part of its daily fishing operations. An ocean-going autoline vessel will cast its lines at about eight knots and propulsion during hauling will be approximately two knots. This may, however, vary from vessel to vessel and according to the weather and the currents.

Fig. Photo of buoy with floater and hauling of line.

Large areas are required for ocean-going autoline (longline) fishing. As mentioned in the introduction, it is important to be aware of how the fishing activities move across new areas on a daily basis. This is an integral part of the operational pattern for this fleet group.

The longline vessels keep moving to new areas as a natural part of their fishing pattern.

The blue lines indicate several rows of lines that have been cast in parallel in a north-south direction (the direction will vary), whereas the yellow lines indicate the subsequent hauling process. From day 1 to day 4 we can clearly see that the vessel has moved across a large area as a natural part of the vessels' fishing pattern.

<text in figure: Day 1, Day 2, Day 3, Day4.

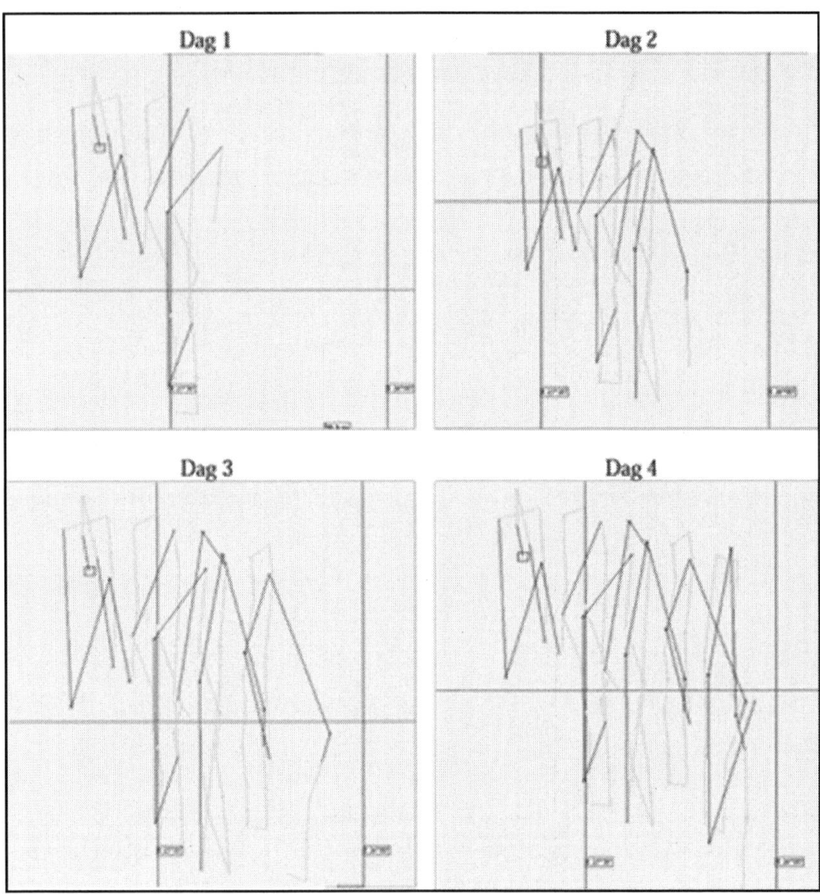

Fig. Plots illustrating how a line vessel moves across different sections from day to day.

MACKEREL JIGGING

Jigging or trolling for mackerel with a hand line is a common fishing method in the south of Norway. Automated mackerel jigging at the commercial scale which is now common is relatively new. The vessels used for this purpose have grown larger over time. Mackerel jigging was previously considered a coastal fishing activity. However, the distances travelled have grown in step with developments in the fishing industry and now include traditional fishing grounds for mackerel in the North Sea.

The jig is rigged as a hand jig with a synthetic fishing line. A weight is attached to the end and several hooks are fastened along the line. The line is dropped into the sea (trailing behind the vessel) and hauled in by means of a mechanical jig.

This type of fishing may take place individually, but is often characterised by a large number of vessels gathering within the same accessible catch area.

The vessels are arranged for fishing by using many jigs simultaneously. This means that the vessel's manoeuvrability is somewhat limited if the vessel is to prevent the hooks from becoming tangled up, despite the fact that the fishing lines do not extend far behind the vessel. The fishing is carried out at relatively low propulsion speeds.

Fig. Photos of vessel jigging for mackerel.

JIGGING

A jig is a hand line with a weight at the end and hooks attached to the line above the weight. The number of hooks varies. Jigging with a hand-held jig still takes place, but this type of fishing is now largely mechanical, using a mechanical jig. Generally speaking, a mechanical jig operates by hauling in some of the line and then letting it out again. This process is repeated until the fish load reaches a certain preset pressure, and then the jig will be hauled to the surface (onto the vessel).

The use of mechanical jigs has the clear advantage that many jigs can be installed on one vessel. It is common to have up to four mechanical jigs on one one-man vessel. As one previously could only use one hand jig per person, mechanical jigs have made jigging much more efficient. Jigging takes place in fjords and proximate coastal areas. Fishing usually takes place whilst the vessel is stationary. However, the vessel will move somewhat as it will drift with the currents and position itself according to the location of schools of fish.

Fig. Photo of vessel cod fishing with mechanical jig.

TRAWLING

A trawl is a tunnel-shaped fishing net which is towed through the water. The water strains out through the mesh entrapping the fish and retaining them in the cod end of the trawl. Historically, trawling is a new method. Trawling started about 100 years ago. Since then there has been significant development in terms of methods and equipment, particularly with regard to the size of the trawl and specialisation according to the type of species one wishes to catch. Thus, different trawls have developed that are better for a particular type of fish than others. The shape and size of the trawl vary significantly. The main factors taken into account when it comes to developing and specialising trawling includes fish behaviour, seabed conditions, selection devices (grating and mesh selection) and the vessel's engine power. The pulling speed during trawling ranges from 1.5 knots to 5 knots. Smaller shrimp trawlers travel at the lowest speed, whereas larger whitefish trawlers and pelagic trawlers travel at the highest speed.

Two main uses of trawls have developed: bottom otter trawl and pelagic trawl. In addition, there is a midway solution referred to as semi-pelagic trawl. In the following, we will describe the characteristics of bottom trawl and pelagic trawl, as well as beam trawl as this type of trawl deviates significantly from the ordinary trawling concept.

The trawling activity pattern using a single bottom otter trawl varies according to several factors such as catch availability, the number of vessels on the field, as well as other circumstances. Below follow two plots which illustrate two different activity patterns for a trawler using a single bottom otter trawl. Both plots have a timeframe of 24 hours.

The left-hand plot vessel trawling a wide area. This may be because the fish is scattered across a large area or because the vessel is pulling the trawl whilst searching for better fishing grounds. The plot to the right the vessel is trawling within a concentrated area, indicating that the fish is located within a concentrated area.

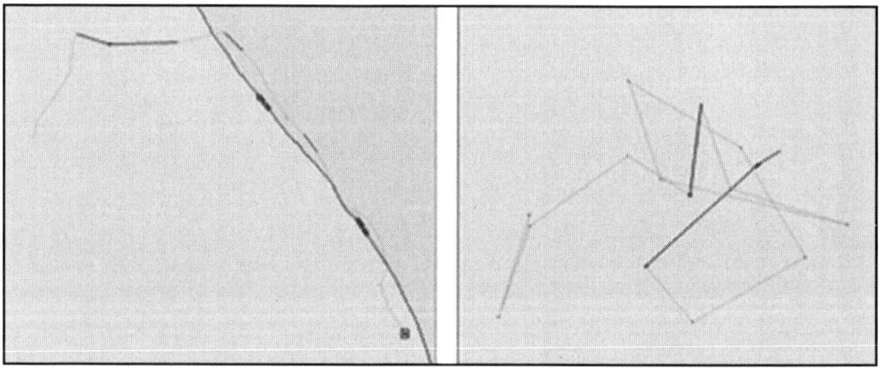

Fig. Example of various trawling activity patterns in a 24-hour period.

BOTTOM OTTER TRAWL

A bottom otter trawl is a trawl which is towed along or close to the seafloor. As the name indicates, a bottom otter trawl will always be in contact with the seabed. The trawl's contact with the seabed is in principle limited to the yaw (bobbin chains, rock hoppers or lighter rigging) and trawl doors (otter boards). The trawl itself consists of a few main components such as the cod end (trawl bag), extension and trawl net. Floats are attached to the upper and foremost part of the trawl net to provide buoyancy and keeping the trawl open. The bottom foremost part of the trawl net ensures that the yaw is in contact with the seafloor.

Two trawl doors, to the front of each side of the trawl, keep the trawl open horizontally when towed by the vessel's trawl wire which is attached to the trawl doors. The largest doors used today (shrimp trawl doors) weigh 4000 – 5500 kilos each and are 10 square metres. The extension of the trawl and how far behind the vessel it is towed depend on the depth and type of trawl. Bottom-otter trawls are further divided into three main groups:

SINGLE BOTTOM OTTER TRAWL

A single bottom otter trawl is characterised by having one single trawl net towed by one vessel. The trawl net is kept open by two trawl doors. The size and extension of the device vary significantly.

Fishing activities with a single bottom otter trawl can be roughly divided into two groups:

1. Trawling for cod, haddock, saithe and shrimp in the entire Norwegian Economic Zone
2. North Sea trawling for blue whiting, Norway pout and sandeel

The former fishing activities are usually carried out by stern trawlers where the catch is hauled on board before the trawl net is emptied. Trawling for blue whiting, Norway pout and sandeel is carried out in more or less the same manner as for pelagic trawl fishing, where the catch is pumped on board while the trawl net is suspended from the sides of the vessel. Some vessels lift the catch on board in smaller lots, so-called "bagging".

Fig. Photo of a stern trawler and illustration of a single bottom trawl.

OTTER TWIN BOTTOM TRAWLS

An otter twin bottom trawl is characterised by having two trawl nets rigged next to each other horizontally and towed by one vessel. The trawl nets are kept open by two trawl doors. There is also a weight between the trawls. The size and extension of the equipment may vary a fair bit. The vessel's manoeuvrability will be somewhat limited compared to that of a single bottom trawl. This type of trawling is sometimes used when trawling for cod, haddock, saithe and shrimp.

Fig. Photo of an otter twin bottom trawl and illustration of a double bottom trawl.

PAIR BOTTOM TRAWL

Pair bottom trawling is characterised by two vessels towing a trawl net between them. The opening of the trawl net is regulated by the distance between the two vessels, thus trawl doors are not used. The size and extent of the equipment vary a fair bit. When the nets are hauled in, both trawl wires are transferred to one of the vessels. Pair bottom trawling is a technique not much used by Norwegian vessels.

<text in ill., from top: towing vessels, trawl warp, weights, sweeplines, floatline, footrope, trawl net>

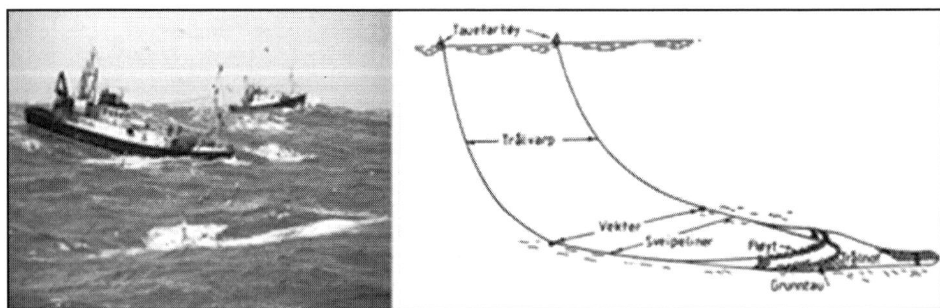

Fig. Photo of a pair bottom trawler team and illustration of a pair bottom trawl.

In addition to the bottom trawling groups described above, there is another type of bottom trawling carried out in the North Sea. This group of trawlers and the trawling device itself is referred to as a beam trawl. The target species for this type of fishing include various types of flatfish.

BEAM TRAWL

Beam trawling is carried out south of 58° North in the North Sea and only by foreign vessels. The fishing method is characterised by a vessel towing two smaller trawls or, more precisely, scraping after the vessel on each side. The equipment consists of a trawl where the doors and the headline have been replaced by a fixed iron construction. Moreover, it is typical for this type of fishing to use short hauls and to trawl at greater speeds than what is common for ordinary bottom trawling. The vessel's manoeuvrability is also better than for ordinary bottom trawling.

Fig. Photo of a typical beam trawl and trawling arrangement towed by the vessel.

PELAGIC TRAWL OR MIDWATER TRAWL

Pelagic trawling is characterised by the fishing gear not coming into contact with the seabed during fishing. Pelagic trawls are used to fish pelagic species (herring, mackerel, blue whiting, greater argentine and capelin). However, a smaller part of the Norwegian fishing fleet also uses trawling to catch herring, mackerel and capelin. It is most common to use seines to catch the fish mentioned above.

The depth where the fishing takes place varies according to where in the water column the fish is located, but it is not common to go deeper than 500-600 meters in Norwegian waters. A pelagic trawler has no yaws of the type commonly used in connection with bottom trawling, but is rigged with a chain on the trawl's groundline. Some trawl types use weights attached to each side of the chain. The weight of these varies a great deal, but for some trawls, it is not uncommon to use weights of 4000 kilos. The extension of the equipment is often greater than for ordinary bottom trawling.

Whereas a bottom trawl often has a top and bottom net, a pelagic trawl is square. This means that there is a top and a bottom panel, as well as two side panels. The function of the doors is to extend the trawl opening horizontally, whereas the weights increase the opening vertically. The trawls can have a vertical opening of 150 metres and a horizontal opening of up to 250 metres. The mesh size of this type of trawl can be up to 256 metres. The mesh size makes it easier to tow the trawl as resistance is lower.

Trawl doors used for pelagic trawling come in different varieties. The size of the doors varies according to the size of the trawl. Doors used for fishing for, for instance, blue whiting will weigh between 2500 and 4000 kilos each and are approximately 13 square metres. Pelagic trawling is characterised by a vessel towing a trawl net not touching the seabed. Special trawl doors designed for pelagic fishing are used. At the same time, weights are used (often chain clusters) to optimise the trawl opening vertically. The depth of a pelagic trawl is adjusted to the length of the trawl wires and the towing power (the speed of the vessel).

Fig. Photo of a pelagic trawler and illustration of a pelagic trawl

DEMERSAL OR SEMI-DEMERSAL TRAWL

Semi-demersal trawling is a combination of the bottom and pelagic trawling methods when a pelagic trawl is lowered down towards the seabed. The trawl can then be towed along near the bottom using two weights, one on each side of the trawl. This method is used to make the most of the catch potential of a pelagic trawl near the seabed, but without towing the trawl itself on the bottom.

<text in figure: warp, semidemersal trawl.

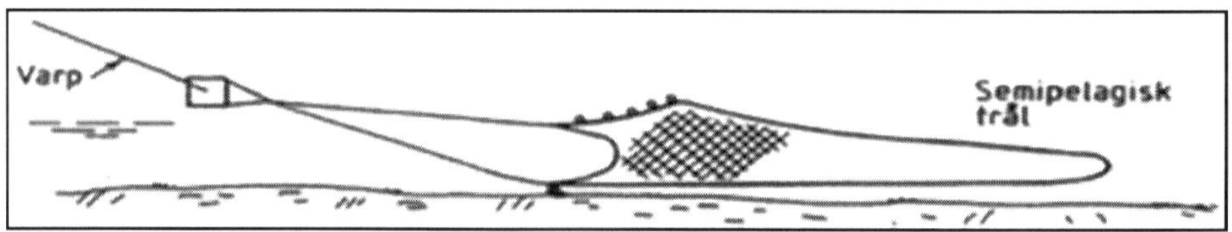

Fig. Illustration of a semidemersal trawl.

PELAGIC PAIR TRAWL

Pelagic pair trawling is usually carried out by smaller vessels and is not very common amongst Norwegian vessels. This method is based on two vessels towing an ordinary pelagic trawl. The opening of the trawl net is regulated by the distance between the two vessels, thus, weights, not trawl doors, are used. The size and extension of the equipment may vary a fair bit. When the net is hauled in, both trawl wires are transferred to one of the vessels.

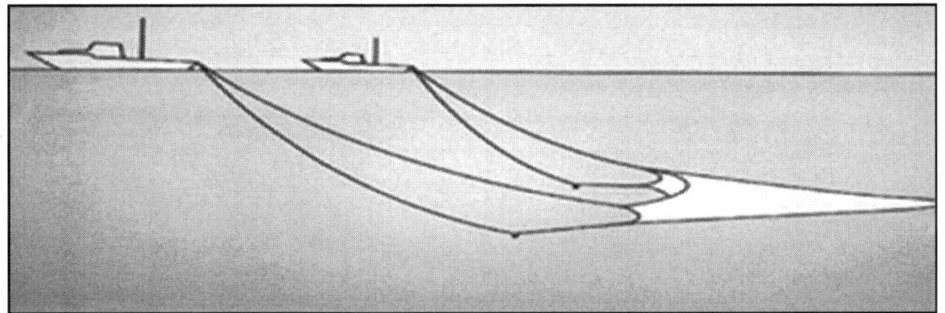

Fig. Illustration of pair trawl with pelagic trawl

SEINES

Seine fishing is a fishing method which shares many of the characteristics of trawling, but which does not employ trawl doors. The difference between trawl and seines is that the trawl is towed through the water whereas seines ideally are hauled through the water.

There are two types of seine fishing:

1. Danish seine
2. Scottish seine

Norwegian fishermen use the Scottish variety. Seine fishing in Norway usually takes place in the north of Norway. The extension and length of the lines vary with the depth and seabed conditions.

DANISH SEINES

This is a type of seine fishing where the vessel is anchored during the catch process. This entails that the vessel is anchored when hauling in the seine after it has been set. As indicated by the name, this is a method frequently used by Danish vessels in the North Sea.

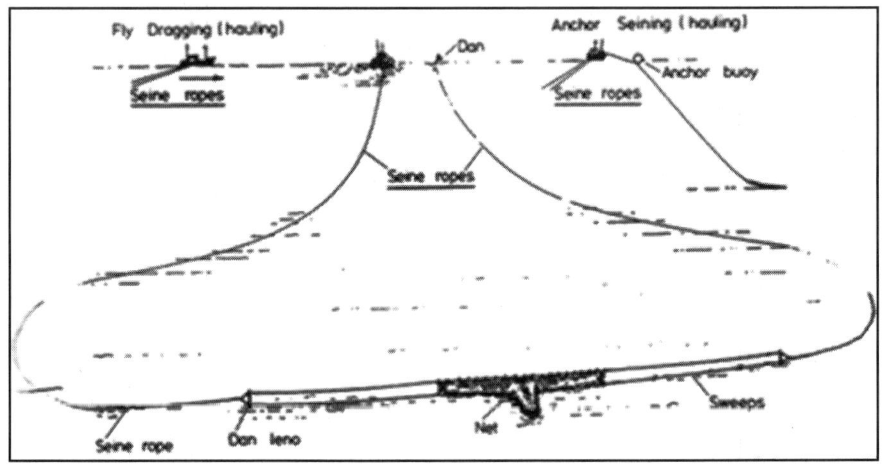

Fig. Illustration of fishing with Danish seine.

The procedure is to cast anchor and drive the rope out in a circle. A drag line is then cast followed by a net wing and the seine. Then the last rope is set (arm) in the same way as the first rope.

Once the line is fully extended, the vessel proceeds at slow speed back to the anchor buoy. The rope will now sink to the bottom and scare the fish towards the middle of the seine opening as the ropes are hauled. The vessel will now remain still and the propulsion of the seine is caused only by the speed of the hauling. The actual extension of the gear will vary according to many conditions, such as depth.

SCOTTISH SEINE

This type of seine fishing does not involve an anchor buoy, but the vessel is kept still by using engine power during the hauling process. The fishing takes place by launching a buoy. Then the vessel drives out a rope and launches the seine before the last piece of rope (arm) is set. Once the seine has been set, the buoy is taken on board and the hauling commences, whilst the vessel retains its position by means of engine power.

<text in figure, clockwise from the right: seine, vessel during the setting phase, marker buoy, rope arm; seine at different stages, rope positions in different stages of hauling, vessel during hauling>

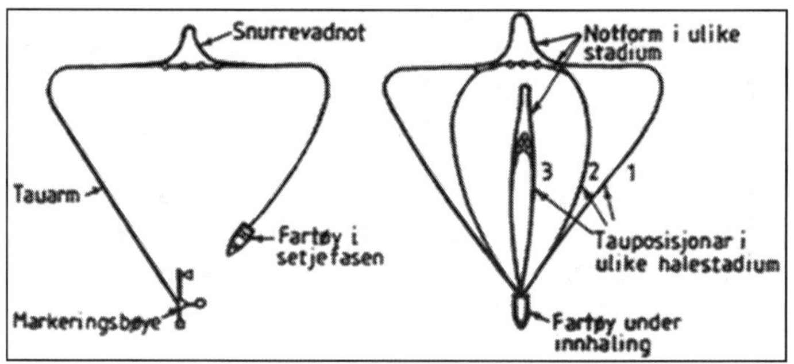

Fig. Illustration of fishing with Scottish seine.

PURSE SEINE

Purse seine fishing has been and still is an important fishing method in Norway. There have been major developments in purse seine fishing. From cotton-thread seines and small vessels using hand power to haul in the seine, to large vessels where the hauling of significantly larger seines has become mechanised.

The modern purse seine vessels are amongst the most efficient fishing vessels in Norway today with regard to the number of different species they catch. We are thinking in this respect of vessels fishing herring, mackerel and capelin. Many of these vessels may also be combined with pelagic trawls and some bottom trawls to catch sandeel and Norway pout.

It is, however, important to be aware of the fact that for seine fishing the total number of vessels is divided into two main groups; the purse seine fleet and the coastal fleet. Shore seine fishing is not considered relevant in this connection. Historically, the coastal fleet consisted of smaller vessels fishing nearer to the coast. Recent developments have, however, entailed that when it comes to seine fishing grounds, there is now little distinction between the purse seine fleet and the coastal fleet.

Seine fishing can be divided into five main phases:

1. Search phase
2. Casting
3. Pulling (closing the net)
4. Hauling the seine
5. Pumping the catch on board

It is important to differentiate between these phases with respect to the vessel's manoeuvrability. During the search phase, the vessel's instruments are used and the vessel has no equipment in the sea. During the casting phase until the pumping phase, the vessel has very little opportunity to move or change course. During these phases the vessel will to a great extent move (drift) in the direction of the current. The casting phase may last from 1 hour (failed cast) to 4-5 hours. However, this will vary according to how much fish is caught and according to the weather and currents. A purse seine is basically a large "net sheet". When pulled, the seine will start to close, entrapping the catch.

<text in fig. Normal total length, approximately 700 m., floaters, purse seine puller, seine ear, purse line, lead line, seine rings, fish bag, buoy, floaters, normal depth approx. 165 m (middle).

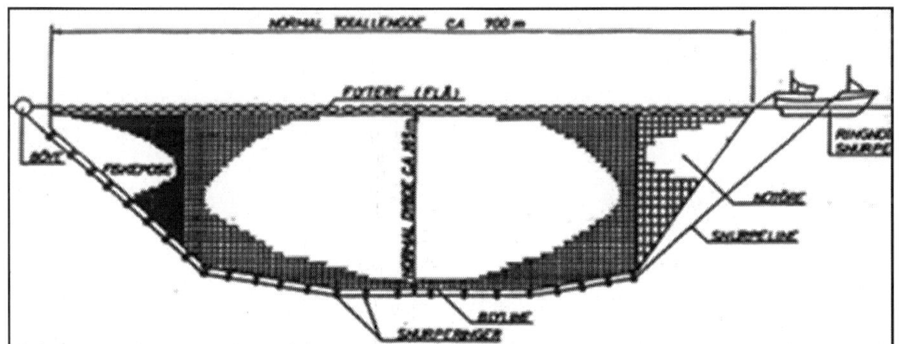

Fig. Illustration of a purse seine

When launching a seine, a drift anchor is cast and the seine released at the same time as the purse line is slackened. The seine is set going in a ring towards starboard and back to the end of the seine which was launched first. The seine is then pulled together at the bottom until the purse line and the purse rings are suspended along the side of the vessel and the catch is entrapped in the seine. The seine is then "dried" using a triplex and passed on to the roller/

power block and stored in a seine bin in the aft part of the vessel. Finally, the catch is compressed so that pumping on board can commence.

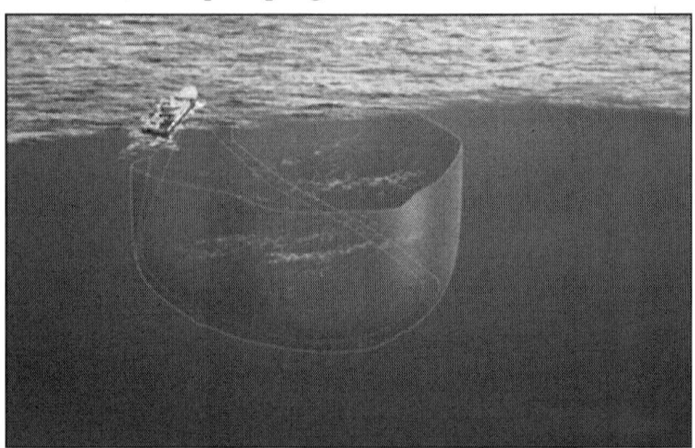

Fig. Illustration of a purse seine cast around a school of fish.

Fig. Photos of a purse seiner and a coastal seiner

STATIC FISHING GEARS

Two types of fishing gears are licensed as static fishing gears: gill and trammel nets, and trap. Gill net is constituted by a series of panels of meshes with a lead foot-rope and a headline with floats. The number of panels can vary.

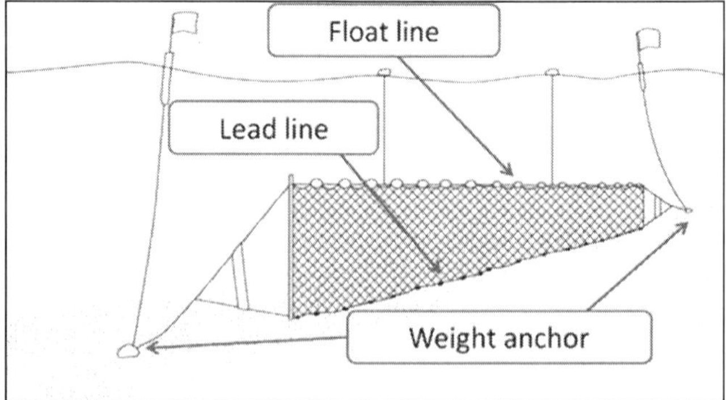

Meshes are uniform in size and shape and therefore highly selective for a particular size class of fish. Fish become snagged by their gill opercula, fins or scales as they try to swim through the net. Gill net can be set at different depth, from the bottom to the surface of the water column.

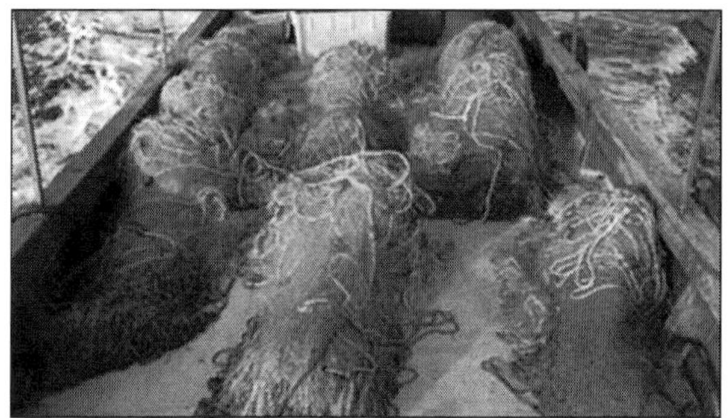

Trammel net differ from gill net because they incorporate three layers of netting: an inner small meshed net between layers of large meshed net. Trammel net is usually set on the seabed. In addition, a combination of the two types of nets is used by vessels of Chioggia's fleet. The net is composed by a trammel net on the upper part, and a gill net on the lower one.

Two types of traps are used by vessels of Chioggia's fleet. A first type, called "nasse", consist in a trap composed by one chamber of non-return. These traps are set in lines, formed by approximately 25 traps, anchored to the bottom, in coastal waters, targeting mainly cuttlefish during their breeding season. A different type of traps, called locally "cogolli", is constituted by several chambers of non-return. Traps are attached to a wall of net that directs the fish to the entrance of the trap. This type of traps is used inside the lagoon, targeting different species.

LONG-LINE

The long-line consists of a length of line, wire or rope, to which baited hooks are attached via shorter lengths of line.

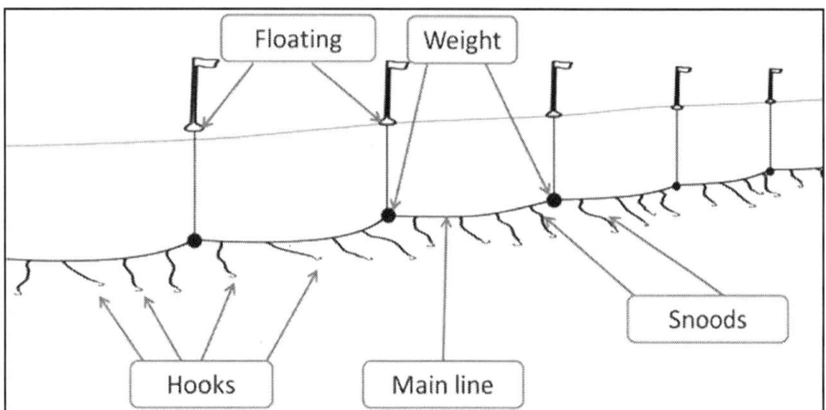

The main line is provided with buoys and weights, whose relative amount determines the depth at which the long-line fishes.

The long-line can be set close to the bottom, anchored at each end and market with buoys at the surface, or close to the surface, drifting or kept attached to the vessel. Long-lines set close to the bottom or at water surface target different species, nektobenthic predators the former, pelagic predators the latter. In the Chioggia's fishing fleet, few vessels use this technique.

FISHING LINE

Fishing line consists of hooks attached to a line and provided with a weight. Fishing line can be attached to a fishing rod. Hooks are usually provided with baits, and can be towed or deployed from a still vessel. This fishing technique is frequently used by recreational fishermen.

TRAWLS

Trawl represents the main fishing technique used by Italian fishing fleets, including the Chioggia's one. The nets can have different shapes, and are held open by buoys, attached to the headline, and a weighted foot-rope or a rigid beam. Trawls can be towed close to the bottom (bottom trawling) or in the water column (pelagic trawling).

-*Otter trawls:* the net includes two wings, held open by otter boards, and a codend.

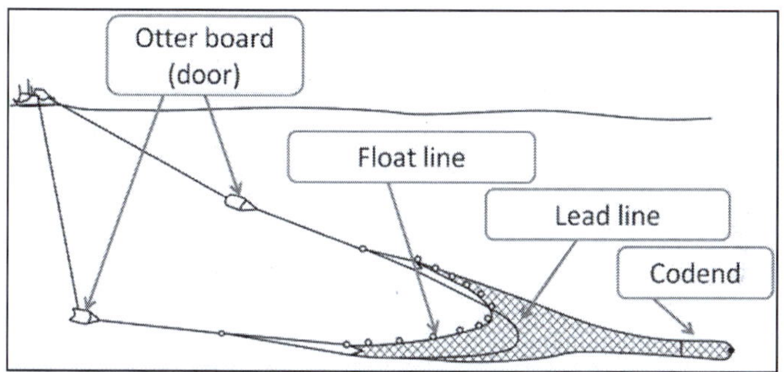

The net can be towed at different distances from the bottom, adding more or less weighted foot-ropes. It usually targets benthic species.

-*Beam trawls:* the net is attached to a rigid beam supported by shoes (four in the Chioggia's vessels). A modified gear ("rapido") is used by fishing vessels from Chioggia.

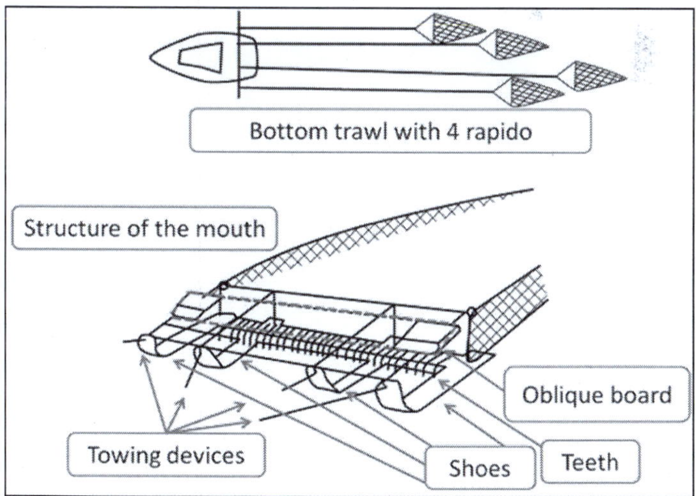

The upper part includes an oblique board,that help in maintaining the net mouth close to the bottom, while lower part is provided with iron teeth (5-7 cm long), that enter into the bottom. A fishing vessel usually tows up to four beam trawls. This technique is frequently used by Chioggia's fishing fleet and targets scallops, great scallop, and flatfish.

Pelagic trawls ("volanti"): this technique consists in two vessels towing together a large net.

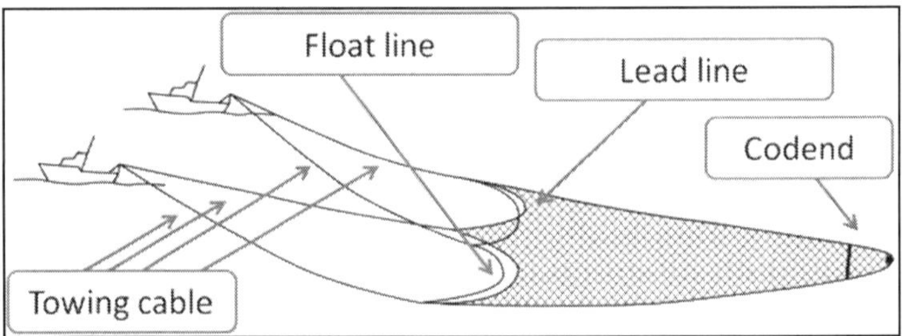

The net mouth is held open thanks to the use of the two vessels, while buoys, attached to the headline, and a weighted foot-rope guarantee the vertical opening. Depending from towing speed and weight of the foot-rope, the net can fish at different depths. Several vessels of the Chioggia's fishing fleet use this technique, targeting mainly pelagic species like anchovy, sardine, and mackerels.

SEINE

This gear represent old fishing technique used on shallow sandy bottoms. The net can have variables lengths (from few to hundreds of meters), and includes a coden and two wings. The net is held open by buoys, attached to the headline, and weights, attached to the foot-rope.

The net is set from the shore or deployed by boat at sea and slowly towed. The towing from the shore implies the work of several fishermen. This fishing gear is today rarely used. Even if some fishing boats from Chioggia have the license for this fishing technique, it is not used any more in the area.

PURSE SEINE

The purse seine consists in a very large net held close to the surface by buoys, attached to the headline, and a weighted foot-rope. It is set encircling a school of fish, usually detected suing a sonar. After the set, the net is closed at the foot-rope and then dragged aboard. A light can be used to attract fish. This technique targets mainly pelagic fish, such as anchovy, sardine, mackerels and tuna. In the Chioggia's fishing fleet, although some vessels have the license, it is not currently employed being substituted by the pelagic trawls.

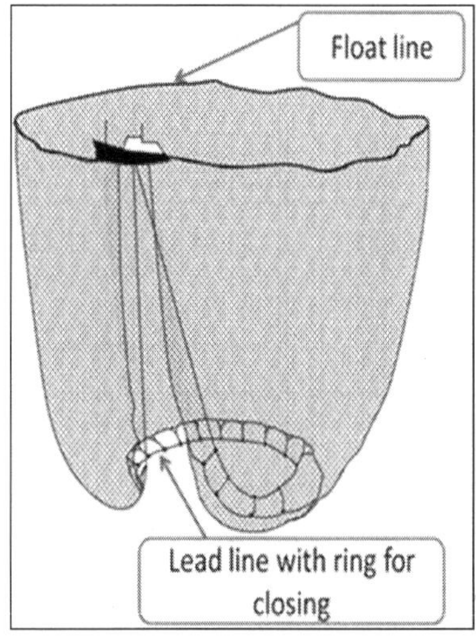

HYDRAULIC DREDGE

Hydraulic dredge ("turbosoffiante") consist of rigid metal "cage", equipped with a blade that enters into the bottom.

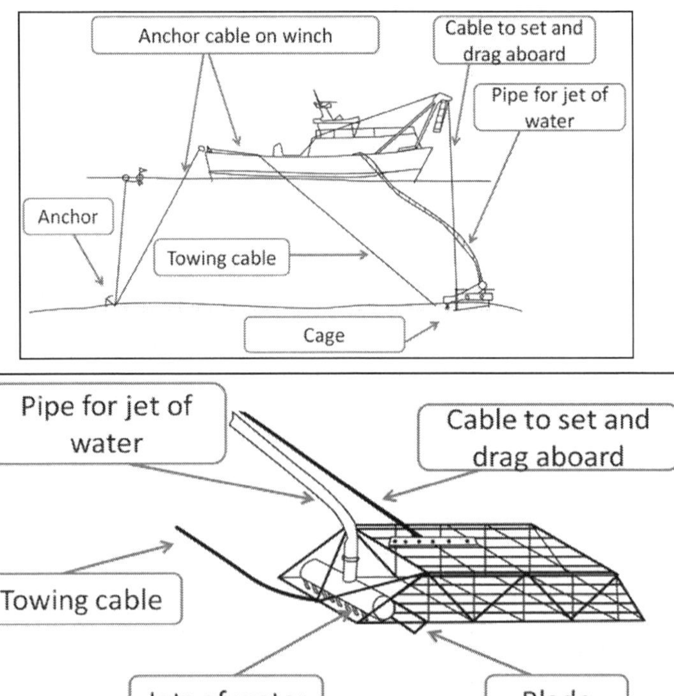

Jets of water, pulled in front of the dredge head, are used to fluidize the sediment and facilitate the penetration into the bottom. The rigid "cage" is deployed from boat bow and towed at reverse. Mesh size of the net can vary, according to the target species, and is regulated by law. Several vessels of the Chioggia's fishing fleet are licensed to use this fishing technique, and the main target species are bivalves, such as striped venus, jackknives, callista, and the peanut worm, sold as baits.

MECHANICAL DREDGE

Mechanical dredge consists in a rigid metal frame and a net. The net is towed either in a straight line or in a circle line, exploiting engine power and anchor resistance. Mechanical dredge is towed on soft bottom, targeting bivalves in particular, for the Chioggia's fishing fleet, clams inside the Venetian Lagoon.

SPEAR

The spear consists of a stick, of variable length, provided with metal teeth at the end. This gear is today employed mainly by recreational fishermen. In Italian water, it is still used to catch swordfish, in a traditional fishing techniques used in Sicily. Only one fishing vessel of the Chioggia's fleet has this license.

6

Development of Fishing Gear and Methods

CHIEF GEAR TYPES IN USE 1900–1940

In the early days of fishing and up to World War II, most commercial fishermen had a market for only a few species of fish. As a result of this, the types of gear and the methods they used were extremely selective. They were selective both as to species and as to size of fish taken. Most fishing boats were single-purpose units fishing practically all the time for one species and for fish above a certain size within that species. Some of the common vessels and gear types of the pre-war era are listed below together with the principal species sought.

Vessel	Gear Type	Fish Sought	Area Fished
Steam Drifter	Drift net	herring	N.E. Atlantic
Long Liner	Baited long lines	cod & halibut	N. Atlantic & N. Pacific
Otter Trawler	Granton trawl	cod & haddock	N. Atlantic
Danish Seiner	Bottom seine	plaice & cod	North Sea
Hand Liner	Hand lines	cod	Lofoten, Greenland, etc.
Pareja Trawler	Pair trawl	hake	Deep water, N.E. Atlantic
Lobster Potter	Lobster creel	lobster	Inshore waters
Ring Netter	2-boat ring net	herring & mackerel	Sheltered waters
Tuna Clipper	Pole and Line	skipjack, bonito, tuna	E. Pacific & C. Atlantic
Purse Seiner	Purse-seine net	sardine & herring	E. Pacific & W. Atlantic

It was relatively easy to control the fishing effort of the above vessels because many of them were working for one kind of fish and their gear in most cases could not be economically used to catch other species. The nearest thing to a multi-species fishery was bottom trawling, but even there a handful of species dominated the catch, especially in the higher latitudes where the bulk of fishing took place.

Fish markets could not cope unexpectedly with less familiar species and there was little incentive to diversify. For most fishermen in the North Atlantic there were only two species that really mattered and these were herring and cod. Under that sort of situation one could fairly easily control the size of fish caught by regulating a minimum mesh size for drift nets and trawls or minimum hook size for long lines.

MAIN GEAR TYPES NOW IN USE

The situation regarding fishing effort has altered radically since the war as a result of enormous advances in fishing technology coupled to a considerable increase in and diversification of fish markets. Today, two types of fishing gear account for the bulk of the fish catch in most countries in temperate regions. These two gear types are the trawl net and the purse seine. But within each type there exists a multitude of variations and sub-types which make simple analysis all but impossible. For instance, we could categorize the main trawls and seines as follows:

Otter Trawls	Trawls Without Otter Boards	Purse Seines
Smooth-ground Bottom Trawls	Beam Trawls	Two-boat Purse Seines:
Bobbin-rigged Otter Trawls	Double Rig Beam Trawls	Menhaden Seines, etc.
High-opening Bottom Trawls	Danish Seines	Lamparas & Ring Nets
Semi-pelagic Otter Trawls	Scottish Seines	Single-boat Purse Seines:
Midwater Trawls	Japanese Bottom Seines	Herring Seines
Single Shrimp Trawls	Bottom Pair Trawls	Mackerel Seines
Double-rig Shrimp Trawls	Midwater Pair Trawls	Capelin Seines
Twin-rig Shrimp Trawls	Bottom Pair Seines	Tuna Seines, etc.

PURSE-SEINE USE AND ADAPTABILITY

Purse seines tend to be used for a single species of fish and the trawls for a variety of species. There are some purse-seine fisheries where a variety of species are taken, but these are mostly associated with light attraction or the use of fish aggregation devices. Salmon purse seines and tuna purse seines are not really suitable for the capture of other species.

However, although a North Sea purse-seine skipper might use his net chiefly to catch herring and/or mackerel, he might just as easily capture schools of sprat, capelin or pilchard. Provided that the other fish were available and that a market for them existed, this could happen. The fishermen's main concern in switching from one species to another would be that the mesh was small enough (to avoid gilling) and that the net was deep or shallow enough for the grounds and the schools of fish in question.

RANGE OF SPECIES CAUGHT IN BOTTOM TRAWLS

With trawl nets the situation is more complex. The net types listed have a basic selectivity related to their design and to their use. But they are all used on the sea bed and for the same general group of species. Taking just six of the bottom trawls used in the North Sea, we might illustrate how their use and effect overlap with each other, as follows:

Bottom Trawl Gear	Type of sea-bed on which it is used:		
	Rough ground	Hard ground	Smooth ground
Bobbin trawl	←———	———→	
Pair trawl		←———→	
Light trawl		←———	———→
Scottish seine		←———	———→
Danish seine			←———→
Nephrops trawl			←———→

Although they are all bottom trawls utilized chiefly for demersal fish, the proportion of different species caught varies as a result of the net design, the techniques used and the grounds fished. Note that the catch composition would still be different if all these trawls were used on the same grounds at the same time.

Table. Relative proportions of different species taken by Trawl Nets in the North Sea - 1980.

Kind of Fish	TYPES OF GEAR					
	Bobbin trawl	Pair trawl	Light trawl	Scottish seine	Danish seine	Nephrops trawl
	(%)	(%)	(%)	(%)	(%)	(%)
Cod	31.0	45.5	24.0	20.0	53.4	8.3
Haddock	32.7	21.9	25.6	44.2	14.8	3.0
Whiting	15.7	12.8	27.2	24.5	3.7	10.8
Saithe	13.2	14.9	3.5	1.1	0.5	2.0
Plaice	0.4	0.35	3.4	3.2	18.7	1.6
Dogfish	1.7	0.9	3.2	1.6	2.6	2.4
Monks	0.8	0.7	3.4	0.9	0.2	3.2
Lemon sole	0.6	0.5	2.1	0.9	2.4	2.3
Skate	0.8	0.25	2.1	0.9	0.3	2.3
Nephrops prawns	-	-	1.8	-	-	59.3
Other species	3.1	2.2	3.7	2.7	3.4	6.5
Total:	100.	100.	100.	100.	100.	100.

From the percentages of different species caught by each of the six bottom drag nets, we can see that their relative effectiveness varies considerably. Taking the closest of the above values for each species and the most variant, we can calculate the difference in relative effectiveness of the gear types as follows:

(times by which one gear may outfish another)					
		Ratio			Ratio
Cod	:	0.03–5.0	Dogfish	:	0.06–3.5
Haddock	:	0.17–8.0	Monks	:	0.14–17
Whiting	:	0.12–7.5	Lemon sole	:	0.14–4.8
Saithe	:	0.12–30.0	Skate	:	0.12–7.0
Plaice	:	0.06–5.5	Prawns	:	300 +

These values reflect the difference in performance during one year only, but are sufficiently diverse to make the point that different types of trawls are selective in different ways.

MIDWATER TRAWL USE

Midwater trawls are used to catch fish concentrated in schools or in layers anywhere from about 10 metres below the surface to about 10 metres above the sea bed. These concentrations consist mostly of fish of one species, and one size or age group.

Herring, mackerel, sprat, anchovy or sardine may be taken by midwater trawls. The fishermen are usually looking for one particular species and they would have a fair idea from the echo-recordings, the place, time, depth and behaviour of the school, what species it consisted of. There is little selectivity in the trawl itself, apart from the size of the bag and codend meshes which are usually small in any case.

Midwater trawls pulled by more powerful vessels may catch faster-swimming, better-quality fish at times than nets towed by smaller boats. But if a layer of fish consists of two species, say herring and mackerel, or herring and pilchard, any midwater trawl towed through them is likely to capture both.

TYPES OF FISHING METHODS

Although there are many methods of catching fish, they fall into three main groups:

1. Catching fish singly or in schools by use of nets or spears
2. Trapping fish in stationary gear, such as fish traps or set nets
3. Attracting fish to get caught on hooks by use of bait, artificial lures or other means such as light

Before you go fishing you must decide what is the best method to use. There are many factors to consider - especially if you are a commercial fisher.

Factors include:

- Where you are fishing
- What species you are fishing for
- The weather and sea conditions
- The cost of the boat, gear and fuel
- The market requirements

NETTING

While there are many types of nets, all rely on the fish getting snared or caught in the net's mesh. Nets are typically long, narrow and flat, weighted at the bottom edge and supported at the top edge by floats. The most common form of netting for recreational fishers is "set" netting. It is also used by commercial fishers to catch fish like flounder and butterfish.

LINING METHODS

The most commonly used lining methods are hand-lines and long-lines. Hand-lines are mainly used by recreational fishers, though they are sometimes used for commercial species, such as southern bluefin tuna. Long-lines consist of a main line running parallel to the bottom, with baited short lines (snoods) attached at intervals. The line is anchored at each end and held at the surface by floats. Long-lines are used to catch high-quality, high-value fish such as snapper.

TROLLING

In trolling, baited hooks or lures are towed behind a boat and fish are pulled aboard when caught. This method is designed to target fast moving surface swimming fish such as tuna, marlin and kingfish.

TRAWLING

Trawling is the most important commercial fishing method in New Zealand, especially for deepsea species. Trawling is used to catch a range of species, for example, orange roughy, hoki, ling, hake and squid. Recreational fishers are not permitted to use trawl nets.

Trawling involves one or two fishing vessels towing a large net. Most New Zealand trawlers are single, rather than pair trawlers. Nets are usually towed for two or three hours at a speed of three or four knots. Nets of both bottom and mid-water trawling are held open by two "doors", which act as paravanes, or underwater kites.

DANISH SEINING

Danish seining is used to encircle, herd and finally trap the fish. A net bag, similar in shape to a trawl bag is operated by a long, weighted rope fixed to each end. The two ropes are used to encircle the fish and also to haul the net in. They are usually operated on the bottom and are used to catch snapper and John dory.

PURSE SEINING

Purse seining is used to catch surface dwelling species such as tuna, mackerels, kahawai and trevally. Aerial spotter planes are usually used to locate the intended catch. The purse seine net is laid in a circle around the school. The net is then "pursed", drawing the bottom closed and entrapping the fish. Purse seining cannot be used by recreational fishers.

BEACH SEINING/DRAG NETTING

Beach seining or drag netting is normally carried out using a length of net and an additional length of warp (rope). The net and warp are laid out from, and back to, the shore and retrieved by hauling on to the shore. The net used is similar to that used for set-netting. Mullet, flatfish, snapper, trevally and crabs are caught this way.

DREDGING

Dredging is used to gather scallops and oysters. To gather scallops, the fishing vessel tows a rigid steel-framed dredge along the sea floor. With oysters, a heavier ring mesh is usually used.

JIGGING

Jigging is a method of catching squid by continuously lowering and retrieving lines from the fishing vessel. Fishing is generally done at night when squid are attracted by powerful lights on the vessel. Jigging is used in preference to trawling when high quality squid is required.

POTS

Rock lobsters and blue cod are caught in pots, usually made of a steel frame, covered with wire mesh. The pot is baited with fish and dropped from the boat on the end of a rope long enough to reach the bottom. The position of the pot is marked with floats so the pot can be easily recovered.

DIVING

Some commercial and recreational fishers dive for paua, scallops and rock lobsters. Paua may only be taken by divers using snorkels, not scuba gear.

MODERN TECHNOLOGY

Today there is more than just luck involved in finding a good fishing spot, especially in commercial fishing. Electronic navigation equipment and sophisticated fish-finding equipment can place the fisher within a few metres of the best fishing areas. On the bridge of a modern deepwater trawler there is millions of dollars worth of electronic equipment, including a colour echo sounder, a colour net recorder, a radar set and a satellite navigation system.

COLOUR ECHO SOUNDER

The colour echo sounder/fish finder is used to find fish and display the depth and contours of the ocean floor. The information is shown on a colour screen. It works through a pulse sent out by a transducer in the hull of the vessel. It strikes the seabed and is reflected back upwards where it is received by the same transducer. Fish also reflect the pulse - it is the air in their swim bladders which makes them visible to the echo sounder. Orange roughy, however, have oil in their swim bladders, and so don't show up very well. This trace from an echo sounder shows possible schools of fish (darker areas).

COLOUR NET RECORDER

This works along side the colour echo sounder. The screen gives specialised information on the net and the fish moving into it, using data from a small echo sounder attached to the net itself. Modern colour net recorders can also measure the width of the net, its fullness and even the temperature at the net mouth. Experienced skippers can tell not only how much fish, but also what kind of fish, they are catching. High tech electronic gear such as this helps commercial fishers operate more efficiently.

RADAR (RADIO DETECTION AND RANGING)

Radar is used to detect objects on the surface up to 30 to 40 km away. Radar works in a similar way to sonar (used in the echo sounder), except it works in air rather than water. Objects the radar identifies show up as "blips" on the screen.

SATELLITE NAVIGATION SYSTEM

Skippers also use a satellite navigation system to determine the position of the vessel. A small screen displays the vessel's position - latitude and longitude - to within a few metres. Accurate measurements can only be made when satellites are above the observer's horizon. At other times the system uses "dead reckoning", based on the ship's course and speed.

Satellites are also increasingly being used at sea to obtain information about weather, sea surface temperature, wave height, wind patterns and the location of fish and other vessels.

THE BEST METHODS ARE NEEDED TO MAXIMIZE THE CATCH AND GET YOUR BIG FISH BEFORE THE SUN GOES DOWN!

There are as many fishing techniques and fish-catching tricks as there are anglers on the water.

Whether you're an experienced angler or just beginning to discover the joys of fishing; whether you're surf casting, bottom fishing, or trolling in deep water; whether you're on the Atlantic, Pacific, or Gulf coasts, this fishing guide of fishing methods and techniques will help you to choose your method, what kind of rod is right for you so you can land more fish.

POLE FISHING

Pole fishing is a highly specialized version of float fishing. Poles provide a level of precision that a rod and reel just can't produce. Just once you learn how to use it, you'll ask yourself why it wasn't sooner.

What makes pole fishing the leading competitive method on fresh water?

- The first and most important reason is bait presentation.
- The short distance between the tip of the pole and the float makes the rig under angler's constant control. Whether holding the float still, moving it left-right with various speeds, slowing it down in a current, or sometimes lifting it up and down, there is always a right way to manipulate the hook bait in order to cause to be it appealing to any nearby fish. Often you just have to copy nature. Just try to match the same look how is the natural forage laying, or moving along the bottom. Also good to use hook bait that matches or imitates native insects or other food items living in the water.
- The second main advantage of pole fishing is accuracy.
- Because we feed ground bait to attract the fish under the pole tip, our hook is constantly around a small concentrated area of feed, i.e. in front of the front of the fish's mouth!
- The third indispensable advantage of the pole is being able to fish with the lightest of tackle (that means a smaller float, thinner line, and/or a smaller hook). It is a fact that lighter tackle always produces more bites, since, obviously, it is less visible to the fish.

Poles can be used to fish shallow or fast rivers, to small ponds or large lakes. They come in different lengths, strengths and designs. They can be used to catch a small fish like Roach, or to tame the hard fighting Carp and even Salmon. Poles allow you to fish with a great level of precision and sensitivity. It is a very popular technique with match fishermen, and when used in the hands of a professional, can result in a huge weight of fish. Poles are big, long and taper to a fine tip.

The line of a float rig is tied onto the end of the pole tip. It can also be set up to attach to the end of a piece of elastic, placed inside the pole tip or attached to the pole tip. The elastic is used to assist in fighting fish. The attached float rig hangs from the pole tip and is literally pushed out over and into the water. The length of line between the float and the pole tip can vary in length for the different types of fishing.

There are a few different techniques that can be used to land fish on a pole:

- Dabbling involves dabbling baits next to stumps, riverbanks, holes in moss beds and other tight spots.
- Sling-shot In order to get a bait or lure under overhead structure like docks or limbs, hold hook by the bend, pull back to put tension on pole and release.

- Strolling is basically manual trolling along a bank that is free from obstructions.
- To tire the fish by making it swim around is the most common method.
- To trick larger fish into swimming toward you and into your landing net. Once you hook the fish, you need to push the pole tip away past the fish, so it is between you and the pole. If you put a bit of pressure on the line, the fish will move away from the pressure, and go towards you.

BAIT CASTING

Bait casting is a style of fishing that relies on the weight of the lure to extend the line into the target area. Bait casting involves a revolving-spool reel (or "free spool") mounted on the topside of the rod. Bait casting is definitely an acquired skill. You can find helpfull information on the How To Fish and Help pages. Once you get the hang of the technique, you will be casting your lures right on target into the places where fish are feeding and hanging out. With bait casting, you can use larger lures (1/2 to 3/4) and cast them for longer distances.

SPIN CASTING

Spin casting is an ideal fishing method for beginning anglers, and the same time professional anglers use it for bass fishing. Spin-casting is easier than bait casting. You can use it to cast both light and heavy lures without tangling or breaking your line. You can use an open-face, closed-face or spin-cast reel for spin casting.

FLY FISHING

To define fly-fishing:

1. An artificial fly is used.
2. The line is what propels the weight-less fly forward, the fly itself has no weight to be cast at any distance.
3. A specific casting technique needs to be performed to cast the line.

With fly fishing, various materials are used to design a very lightweight lure called a fly. Fish think the fly is an insect and they take the bait on, or just above, the surface of the water. It takes a little practice, but fly fishing is a pure and exciting way to fish.

Unlike other casting methods, fly fishing can be thought of as a method of casting line rather than lure. Non-fly fishing methods rely on a lure's weight to pull line from the reel during the forward motion of a cast. By design, a fly is too light to be cast, and thus simply follows the unfurling of a properly cast fly line, which is heavier and casts easier than lines used in other types of fishing.

The angler normally holds the fly rod in the dominant hand and manipulates the line with the other close to the reel, pulling line out in small increments as the energy in the line, generated from backward and forward motions, increases.

Steelhead Trout and Salmon techniques Floating line/indicator nymphing is a great method for smaller rivers where the fish are in less than 6 feet of water. This method is effective when sight fishing to fish holding in shallow water; allowing you the option of placing your fly right in front of a resting fish. Usually the best to use 8-9 wt rods and over-lined reels with a heavier weight floating line.

Spey fishing is the most popular method. Swing flies on long spey rods and pointed leaders, trying to attract a fresh run steelhead to take your offering.

Trout and Smallmouth Bass techniques Floating lines. Light rods, light floating lines, tapered leaders, small flies, and rising trout. 90% of a trout's looking for food below the surface. 4-6 wt fly rods and light reels holding your floating line. Sinking lines. Cast to the banks and structure while presenting your fly through the water. You will have tremendous, aggressive strikes! 5-7 wt fly rod and reels are loaded with either full sinking lines or sink tips.

DOBU OR KOROGASHI FISHING

Dobu is Japanese traditional freshwater fly fishing method, "Old Schools" since the feudal time, using group of nymph-like wet flies in a sinker rig. It was popular game in the medieval time, and mostly enjoyed in western Japan circling Kyoto where craftmen worked the fly patterns into the state of art. Uses long handrod from 9 to 11m (27-33 feet), 4-5m leader with 3-4 droppers, and tenbin boom with sinker on one end and another dropper on the other end. Angler reaches out the rig into the point on side, gently soak it to designated depth, and scan until the rig reaches the most downstream point drawing arc. Mostly used it to catch ayu, dace, and chub.

TENKARA FISHING

Tenkara is Japanese traditional freshwater fly fishing method using single soft hackle fly to catch mountain creek trouts such as yamame and iwana. The former became artifacts for it was enjoyed by swordsman class, and the latter became craftwork of remote site fishermen. Primarily used for small stream trout fishing, tenkara today is one of the most popular methods of angling among fresh-water fisherman in Japan. It is particularly effective for fishing pocket water and faster flowing streams. Some of the main advantages are its elegant simplicity, the very delicate presentations with the light line, the ability to hold the line off the water and a fly in place on the other side of a current, precise casting, and greater control of the fly.

Traditional tenkara used bamboo hand rod, horse tail for taper line, and silk for tippet. Fly patterns were soft hackle which cover mostly beaneath the surface film. Modern tenkara use glassfiber or high-grade carbon-fiber, extremely light and strong hand rods the most portable of fly-fishing rods in 3:7 reflex for taper line and 5:5 reflex for level line. Lines are now made out of fuloro carbon and fly patterns welcome western dry fly patterns as well. They also have a handle similar to fly-fishing rods that can be made of wood (the more prized rods) or cork.

As in fly-fishing, it is the tenkara line that propels the weightless fly forward. Most common rig uses running line (tapered or level) (tapered furled line is twisted monofilament) between 3.9-7m and tie a 30cm-1m 4lb test leader (tippet) to the end of the line. The length of running line is deterimined by the size of the stream, and in the deepest of small stream, leader is directly tied to the tip of the rod to execute dapping. The main advantage of furled lines is the delicate presentation and ease of casting. They are easier to cast against the wind. A regular fly-fishing tippet is used to connect the fly to the line (which is too thick to tie directly to the fly). Artificial flies are used in tenkara fly-fishing. These are tied with thread, feathers and sometimes fur as in western fly-fishing. Traditionally used a special reverse hackle wet-fly known as "kebari" in Japan.

Casting requires a slower and shorter stroke to completely turn the tippet over. A snap cast better to use when wanting to change direction of the fly without false-casting, while an under-hand cast is good for getting under a tree. Tenkara is very well suited for: fishing a dry-fly (virtually drag-free due to the supple line and little line in the water), dapping a fly on a pool or holding it on an eddy, Czech-nymphing and high-sticking, playing a soft-hackle wet fly, and manipulating the traditional reverse hackle flies.

The basis is the action which is the same as fly fishing. The grip places a single finger on the top of the handle. This is good for rod control. There are two kinds of basic casts, the overhead casts and the backhander casts. The back-cast throws up the line on the head and stops it in the position at 1 o'clock. After Line is stretched behind, it is pushed forward. When not learning this timing with the body, the level line doesn't stretch. The fore-cast is made by pressing forward with the whole body. The rod position is stopped about 10 o'clock. The fly returns a few(5cm) arm immediately before falling to the surface of the water. It should land softly on the water. It lands softly on the water when excluding power from the line.

AYU FISHING

Ayu fishing is one of the several narrowly defined styles of fishing in Japan. It was practiced by Samurai as long as 430 years ago. It uses very long rods (5 - 7 meters) and flies but not fly-casting is required. Ayu fishing originated at least 430 years ago when anglers discovered they could dress their flies with pieces of fabric and use those to fool the fish. The art became more refined as the samurai, who were forbidden to practice martial arts and sword fighting in the Edo period, found this type of fishing to be a good substitute for their training: the rod being a substitute to the sword, and walking on the rocks of a small stream good leg and balance training. "Only the samurai were permitted to fish. So, the samurai wo enjoyed ayu fishing would take sewing needles and bend them themselves, and make their own flies by hand.

Ayu fishing may be done with flies or with a decoy fish. As ayu fish are very territorial, they usually attack the live decoy fish used as bait.

DAPPING FISHING

Dapping can be considered a style of fly-fishing, at least by our definitions, but tenkara fly-fishing is not dapping. Dapping ranges from the ancient method of simply touching the fly up and down on the surface of the water, to the modern use of blowline by which the wind carries the fly to fish. Dapping may be seen as but one of the techniques that can be practiced with a tenkara rod, just as it may with a western fly-rod, though a short rod would make it more difficult. Dapping rods are commonly much longer, as long as 40ft, and modern "dappers" could use reels as well. A reel must be with a fast retrieve to bring in a fish and keep the blowline from getting wet. Dapping casting" techniques: ones only uses the wind to blow on the line, then drops the fly in place and maybe imparts some life to it.

STILL FISHING

The simplest of fishing methods can also be the most effective. Still-fishing is a matter of putting your bait in the water and waiting for a fish to find it. Depending on water depth and what you're trying to catch, you may want to still-fish near the surface, at a mid-water depth, or right down on the bottom. Using a float, or bobber, makes it easy to fish near the surface, or you can add sinkers to your line to fish deeper. Still fishing is a versatile way to go. You can do it from a pier, a bridge, an anchored boat or from shore. You can still fish on the bottom or off the bottom in ponds, lakes, rivers and streams for a variety of species. And you can still fish during most seasons and during any part of the day. You have to be very patient. You need to wait for the fish to bite.

DRIFT FISHING

Drift fishing allows you to fish over a variety of habitats as your boat drifts with the currents or wind movement. You can drift fish on the bottom or change the depth with a bobber or float. Natural baits work best. But jigs, lures and artificial flies will produce good results, too. You can drift fish on ponds, lakes, rivers and streams any time of the day and year.

LIVE LINING

Your line is "live" when your boat is anchored in a flowing body of water like a river or stream. Use live or prepared baits and keep them on or just off the bottom. Live lining off the bottom allows your line to drift with the current through holes and rocks where the fish may be holding.

LONGLINES

Longlining attracts fish with a central fishing line that ranges from one to more than 50 miles (80 km) long and carry thousands of hooks. This central line is strung with smaller lines of baited hooks, which dangle at spaced intervals. Baited hooks are attached to the longline by short lines called snoods that hang off the mainline. After leaving the line to "soak" for a time to attract fish, longliners return to haul in their catch. Longlines are set horizontally either on the ocean floor (demersal longlines) or near the surface of the water (pelagic longlines). Demersal- or "bottom"-longliners float their hooks just off the seafloor to catch fish that live on or near the bottom, such as cod or halibut. Pelagic longliners used to catch open ocean fish, such as tuna and swordfish.

DROPLINE

Droplines are similar to longlines but are set vertically either down underwater cliffs or just in the water column. They have a weight at the bottom, a series of hooks attached to snoods, and a float at the top of the line. They are not usually as long as longlines and don't have as many hooks.

CHUMMING

To attract fish or get them biting again, you can throw "chum" into the water where you're fishing. You can use ground-up bait fish, canned sweet corn, dead minnows in a coffee can (for ice fishing), pet food, even breakfast cereal. Or stir up some natural chum by scraping the bottom with a boat oar. Be sure not to over-chum. You want to get them interested in feeding; you do not want to stuff them before they get a chance to go after your hook. Chumming is not legal in all states. Check local fishing regulations.

BOTTOM BOUNCING

Bottom Bouncing is done from a drifting or trolling boat, and it's a great way to attract or locate fish during most seasons and times of day. Use a buck tail jig or natural bait and drag it along the bottom. The dragging motion causes the lure to bounce along stirring up small clouds of sand or mud. After a few strikes with bottom bouncing, you can drop anchor and apply other methods to hook the particular kind of species you've attracted.

TROLLING OR TRAWLING

Most trolling is done using a small electric motor that moves the boat quietly through the water so fish aren't spooked. But you can also troll by towing a lure while walking along the edge of a shoreline, bridge or pier. The speed of the boat determines the depth of your bait. And the depth of the bait is determined by the species of fish you're trying to catch. Use a spinning reel or a bait caster for trolling.

Trawling is one of the most common methods of fishing in the world. Trawling involves towing one or more trawl nets behind a boat or in between two boats, either through the water column or along the ocean's floor. Trawl nets are usually shaped like a cone or funnel with a wide opening to catch fish or crustaceans and a narrow closed end called a cod-end. Trawls can be used in water of various depths down to around 3000m, and nets differ by their mesh size. Midwater trawlers catch faster swimming schooling fish such as sardines. Bottom trawlers catch fish that live on or near the seafloor, such as cod, flounder and shrimp.

PURSE SEINING

A purse seine is a large wall of netting that encircles a school of fish. Fishermen pull the bottom of the netting closed (like a drawstring purse), herding the fish into the center. Purse seiners either haul the net aboard or bring it alongside the boat to scoop out the fish with smaller nets. This method used to catch schooling fish, such as sardines, or fish that gather to spawn, like squid. The most popular fish caught by purse seines are tuna used for canning.

JIGGING

Jig fishing is popular and challenging. Why? Because the person fishing is creating the action that attracts, or doesn't attract, the particular type of fish he or she is trying to catch. Here's how it works. Cast out and let your jig hook sink to the bottom.

Then use your rod tip to raise the bait about a foot off the bottom. Then let it drop back to the bottom. You can jig up and down, side to side or up and down and sideways. Jig rigs come in all sizes, shapes and colors, and can be used with or without live bait.

JIG AND WORM

Attach a worm to your jig hook and use it to bottom hop or sweep through your target area. To bottom hop, cast to the target and let the jig sink. Then reel in slowly, twitching the rod with every third or fourth turn of your reel. To sweep, cast to the target and drag the jig parallel to the bottom while reeling with a fairly tight line. Slow and steady gets the fish when you're sweeping with a jig and worm.

ICE FISHING

Fishing through a three-foot hole in the ice? Yup. It's a unique way to catch multiple species of northern, fresh-water fish. And thanks to advancements in garment design, portable fish houses and fish locating devices, it's becoming more and more popular every day. One- to three-foot rods are most often used and simple reels hold the line. You can also ice fish with tip-ups. When a fish hits your tip-up gear, it releases a lever that raises a flag or rings a bell.

FISHING FROM BOATS

Big or small, motorized or outfitted with paddles, a boat simply allows you to cover more water. But for most folks, a boat simply makes fishing a lot more fun. The boat you choose should be based on where you want to fish and what kind of fish you're stalking.

CATCH AND RELEASE

Catch and release was a way to reduce the cost of stocking hatchery-raised Trout In Michigan. Since then, conservationists and sport fisherman alike have promoted catch and release to ensure sustainability and avoid over fishing.

Proper catch and release fishing techniques also reduce fish fighting and handling times and help avoid damage to fish skin, scales and slime layers—damage that can leave fish vulnerable to fungal skin infections.

TROPICAL AND MULTI-SPECIES FISHERIES

RANGE OF SPECIES AND SIZES OF FISH

In a multi-species fishery like the North Sea bottom trawl fishery, three species comprise over 75 percent of the catch. In many tropical fisheries one would count ten or twenty most prevalent species before totalling 75 percent of the total trawl catch.

If it is difficult in the North Atlantic to manage stocks of cod, haddock and whiting mingled together and caught together, how much more complex is it to manage three or four times as many species. This is a major dilemma facing fishery managers in the warm water countries.

In most developing countries all fish caught can be marketed, whether large or small, mature or immature. There is little incentive for any fisherman to throw juvenile fish back into the sea. The range in size and shape of fishes caught is such that any mesh restriction to limit capture of immature fisih of one species is bound to result in the loss of many mature fish of another species.

SELECTIVE TYPES OF GEAR

However, of fishing gear types and methods used in the tropics, some are extremely species selective. Even in waters where other species abound, their use does not create a complex management problem. The main selective gear types are listed below.

Selective Types Of Gear		
Gear Type	**Species Sought**	**Other Fish Taken**
Pole and line	skipjack & other surface tunas	almost none
Midwater trawl	mackerel or sardine	very few
Squid jig	squid	very few
Drift net	small tuna, salmon, barracuda	occasional shark
Tuna long line	large tuna	billfish & some shark
Scoop net	fry of milkfish, shrimp, grey mullet	very few
Trolling line	tuna species, jacks, mackerel, dolphinfish	no undesirable fish
Shellfish dredge	clams, scallops, oysters	starfish, sea urchin, flatfish
Tuna purse seine	tuna species	porpoise occasionally
Saury lift net	saury	almost none

LESS SELECTIVE GEAR TYPES

Fishing gear and methods which are less selective include most of the demersal fish trawls, the seines and lift nets used in conjunction with light attraction or FADs, and many of the artisanal traps, lines and tangle nets. They are listed below.

Less Selective Gear		
Gear Type	**Main Species Sought**	**Less Desired Species Caught**
Lift net*	sardine, mackerel, anchovy	bait fish, juveniles
Hand line	snapper, grouper, bream	shark, eels, ray
Bottom trawl net	demersal species & shrimp	trash fish, dollarfish, juveniles
Trap net	most edible species	shark, eels, stonefish
Purse seine*	tuna, sardine, mackerel, anchovy	small tuna & juveniles
Bottom Gill net	most edible fish & shrimp	ray, shark, jellyfish
Beach seine	any edible species	starfish, jellyfish
Bottom long line	larger demersal fish	eels, dogfish

Note:

* Used with light attraction or FADs

ANALYSIS OF MULTISPECIES TROPICAL FISHERIES

Multi-species fisheries and the interactions within them are complex and are not clearly understood.

Despite being the focus of a considerable literature, no clearly defined management tools/strategies are currently available. Fishing effects on multi-species fish resources may be detected as changes in the relative abundance of species (species composition) in the stock, linked to fishing intensity, the relative catchabilities of the species, and the level of interaction (e.g. competition and predation) between them.

Fishing removes top predators and is postulated to lead to increase in prey; however, observations supporting these predicted effects are limited. Understanding multispecies responses to fishing is necessary in order for managers to make informed decisions about appropriate resource assessment models, and to develop a suitable management strategy for them.

APPROACH

The project aimed to assess fishing effects, derive biological management guidelines, and describe minimum data requirements for demersal bank and deep slope reef fisheries - relatively simple multi-species examples but with widespread applicability.

To describe observed effects of fishing at the community level, data typically collected by developing country fisheries institutions from artisanal fisheries were analysed for case study fisheries in Tonga (deep slope) and the Indian Ocean (banks). Available local literature described the socio-economic context of each fishery.

To examine expected fishing effects and to develop guidelines for management, theoretical studies were performed using a Multispecies Interactive Dynamic Age Structured simulation model (MIDAS). Single and aggregate single species approaches, with length based methods were employed. Input parameters for modelling (e.g. abundance and population demographic estimates) were derived from case studies.

FINDINGS

A Multispecies Interactive Dynamic Age Structured simulation model (MIDAS) was developed during the project, which simulates a number of stocks being fished by a number of gear types and will determine equilibrium and transient dynamics of the stocks.

No evidence for prey release or other multispecies interactions was found from either case-study or modelling analyses, but species composition changes due to technical interactions were significant. Prey release was only predicted from 5 years onwards into a fishery, but its magnitude was less than variation typically observed in available data indicating that it would be undetectable.

These results suggest that single, and aggregate single species models were adequate to derive management advice and also that data need only be collected on the most important species and guilds of others. However, biological data collection on identified key species and improved data collection in relation to technological changes in the fishery were required.

Management simulations have resulted in guidelines for selecting a suitable target fishing mortality for species where length at maturity is unknown and length at first capture cannot be controlled. The guidelines suggest a means of selecting the most important and vulnerable species for analysis, and a method for setting overall effort limits for a multispecies fishery, taking into account different targeting practices and conservation trade-offs.

GEAR SELECTIVITY

A key definition of selective fishing refers to a fishing method's ability to target and capture organisms by size and species during the fishing operation allowing non-targets to be avoided or released unharmed.

In a fishing area, a range of fish and other species and sizes normally occur together. When encountered by a fishing gear, they will be captured at different rates, depending on the gear design and its mode of operation. Various gears are known to be more selective in this regard than others. A purse seine has a poor selective performance, as it will catch most of the surrounded individuals but it is used on specific schools that are actively sought and identified before being surrounded, limiting the problem to that school and its co-schooling species. No gear is known to be one hundred percent selective for a given species or size range. Most gear and methods fall into a group that has some selective performance, and their ability to select targets can be altered through modification of design and operation methods.

The catch in many fisheries thus consists of a mixture of target and non-target species. What does or does not comprise targets depends to a large extent on the market and whether there are regulations in place prohibiting capture of species or certain sizes of organisms. Non-targets are often synonymous with bycatch, a concept defined differently by various people. A generic definition often used is that bycatch refers to the accidental capture of any species, of any size or sex that is not specifically aimed at during fishing activity. Fishermen may aim at a mix of species. Bycatch, as a choice of the fisher, may be retained or discarded, and is mostly dead when discarded.

When bycatch is landed, the issue, as for any other landing, is the capability to collect information, assess the stock, and manage the multispecies mix with all the inherent difficulties.

When bycatch is discarded, the implications are of biological, ecological, socio-economic and ethical nature:

1. *Biological:* As discards are most often not registered, the status of the populations concerned might be unknown and the risk of extinction may be high;
2. *Ecological:* The depletion of the species concerned may lead to important changes in the trophic chain and ecosystem structure (a problem existing also with the target species). In addition, the development of a dependency of some scavengers (e.g. sea birds) on discards may create a risk for the animals;
3. *Economic:* In addition to the waste of fuel (for catching it) and of manpower (to sort it) the mortality of the discarded populations may have repercussions on the neighbouring fisheries, e.g. because juveniles of their targets are killed and discarded;
4. *Ethical:* As in hunting, the public does not consider ethical to kill and

throw away harmless animals. Discards are then consider as a waste of otherwise useful living creatures.

The adoption of the ecosystem and precautionary approach, and the Convention on Biological Diversity have increased the necessity of regulators to pay attention to these factors that, most frequently, form the basis for their regulatory actions.

The total global discard is difficult to estimate. An early FAO estimate indicated an average annual discard of about 27 million tonnes in 1980-1990, a figure revised later on to about 20 million tonnes. The more recent FAO estimate for 1990-2000 is about 7 million tonnes. Even if comparisons are not completely warranted, the difference indicates a substantial decrease in discards, due in part to improved selectivity but also in part to a more intensive complete use of the catch.

Besides bycatch of fish and other food species, other animals might incidentally be captured with fishing gears, including various species of marine mammals, turtles and seabirds. The impact on their population size may not always be known but public concerns make it necessary to reduce such bycatch.

POSSIBLE SOLUTIONS

Improved selectivity can be achieved in different ways, by modifying the gear design and/or operation and by using alternative fishing gears. In trawls and gillnets mesh size is a well- known measure to regulate the size of captured organisms. For mobile gears, like trawls and seines, improved selectivity can also be achieved by using square meshes in the codends and by inserting filtering grids in front of it. Successful separation of targets and non-targets species can also be achieved by using grid devices.

The principles for such selectivity are detailed in the referenced documents.

Successful technical solutions have also been found to reduce capture of non-fish species like mammals, turtles and seabirds. The capture of dolphins in the purse seine fishery for tuna has been reduced to an insignificant level by using a combination of technical changes, rescue techniques, education of fishers and management actions. Bycatch of turtles in the tropical shrimp fishery can be avoided by using a turtle excluder device (TED), which is a rigid or soft (netting) structure inserted in the aft part of a trawl, in front of the codend proper. Incidental capture of seabirds in longline fisheries can be significantly reduced by underwater setting of the line, night setting and scaring the seabirds away from the baited hooks.

RECENT ACTION

At international level the most prominent action to reduce bycatch during fishing is the ban on large size driftnetting on the high seas (UN General Assembly resolution 1990). The FAO Code of Conduct for Responsible Fisheries of 1995 is a major international document addressing the problem with selective fishing. The most recent international contribution is the International Plan of Action for Reducing Incidental Catch of Seabirds in Longline Fisheries adopted by the FAO Committee on Fisheries in 1999. FAO has recently developed a project (with GEF funding), to enhance the capability of developing countries to reduce environmental impact of tropical shrimp trawling through the introduction of bycatch reduction technologies and change of management.

At national and regional levels regulations, minimizing unwanted bycatch and discards are being introduced. These include minimum mesh sizes, mandatory use of selective devices like TEDs and sorting grids, areas closed for fishing, etc. Many countries are putting a great deal of effort into research and development aimed at improving the selectivity of fishing gear and methods that presently capture much of the unwanted catch.

OUTLOOK

Concerns about bycatch are expected to increase in the future, and therefore focus will continue to be on the development of selective technologies. It is expected that better knowledge of organism behaviour and technological developments will increase the success of such developments.

SELECTIVITY OF TRAWLS BY DESIGN

Trawl nets and bottom seines possess an initial selectiveness owing to their particular design and mode of operation. A mesh size limitation would vary considerably in effect from one type of trawl net to another. It is important therefore to consider gear selectivity and performance before we begin to think of mesh sizes and their effect.

Take a herring vinge trawl for example and two other bottom other trawls like the granton trawl and the shrimp trawl. All use otter boards, all are on the sea bed throughout the duration of the tow and all have the same basic structure of wings, square, bellies, bag and cod end. But their performances differ as much as those of a racing car, a truck and a tractor. All three nets could be fishing in the same area, yet the vinge trawl might take only herring, the granton trawl only demersal fish and the shrimp trawl primarily shrimp or prawn.

Mesh size has little to do with this species selection. It is a function of other aspects of gear performance - speed of tow, headline height, ground contact, flow of water through the body of the net, otter door spread, length of ground-cables, and so on.

The vinge trawl would have the smallest mesh in the cod end, yet it would take no cod, haddock or shrimp. The shrimp or prawn trawl would have a smaller mesh than the granton trawl yet it would capture only a small proportion of the bigger fish the granton trawl would take. To imagine that in these cases the size of mesh controls the size of fish caught would be as simplistic as to think that the respective speeds of the racing car, truck and tractor were determined by the size of their wheels.

In contrast to the trawls designed to capture one species or group of species, there are the multi-species trawls or combination trawls which take a great variety of fish. The North Sea prawn and fish trawl is a combination net designed to capture Nephrops norvegicus or Norway prawns, plus demersal fish like cod, haddock, skate, monks, plaice and lemon sole. Most bottom trawls in the tropics and sub-tropics are multi-species trawls taking fish which vary greatly from each other in size and shape.

There is no common selectivity factor and no common minimum size or length for the various species. For these fisheries, a mesh regulation is at best an inadequate means of reducing juvenile mortality. It can be aimed only at the smallest of the main commercial species.

SELECTIVITY BY AREA SWEPT

Many types of bottom gear depend on the area of sea-bed swept for their effectiveness in fish capture. Ropes, wires, cables and bridles are used to herd fish in the direction of the net. It would be uneconomic in terms of gear and power necessary to sweep such large areas with walls of netting.

Several species of fish respond to herding by ropes, and possibly just as many do not. Generally speaking, the demersal fishes of temperate waters are vulnerable to this type of gear while crustaceans and most pelagic fishes are not. There are several technical, environmental and behavioural reasons for this. Fish respond, not just to the approach of ropes and cables but also to the noise or vibration of the otter boards, and to the clouds of mud the boards leave behind. When the angle and the speed are such that the expanding trail of mud-cloud just touches the ground cable or sweep lines, then for many demersal fishes the herding effect is maximized; this applies only to trawls utilizing otter boards. Danish seiners and pair trawlers are able to herd fish effectively without disturbing the sea bed unduly.

The species that do not respond to herding by ropes include practically all of the shrimps and prawns, and most of the clupeids and mackerels. Trawl nets designed to capture shrimp have a wide but low net mouth and extremely short cables or bridles. Trawl nets for clupeids and mackerels have very large net mouth openings. To make the net mouth as big as possible without creating too heavy a towing load, large meshes are used. These meshes are big enough for the fish to swim through with ease, but this does not happen if the net is properly constructed. So pelagic species while they do not respond much to ropes or wires, will be guided by sheets of tight netting even if the mesh sizes are very large.

Examples of gear types which depend primarily on sweeping a wide area of the sea bed are listed below:

Bottom seines: (beach seines, Danish seines, Canadian pair seines)

Pair trawls: (bottom pair trawls, utilizing long warps and cables)

Otter trawls with Vigneron Dahl gear: (long cables inserted between the otter boards and the net or bridles)

Fishing gear which does not rely on herding fish from a wide area on each side of the trawl net is mainly used to catch shellfish. Some common examples are listed below:

Shrimp trawls	:	double rig and single rig
Nephrops trawls	:	except combination nephrops/fish trawls
Lobster trawls	:	used offshore in deep water, N.E. USA
Beam trawls	:	for flatfish, shrimp and sole
Dredges	:	for scallops, clams and oysters

Midwater trawls are used almost exclusively on schooling fish or scattered layers of schooling fish, and skippers always aim to tow the net directly into or through a school or layer. There may be some scaring effect from otter boards, weights, bridles or kites, but it can hardly be considered a herding effect.

EFFECT OF TICKLER CHAINS AND GROUND ROPES

The ground rope of a trawl net has an effect on the amount or proportion of flatfish and shellfish that may be caught. A heavy groundrope will tend to scrape tightly on the sea bed while a lighter one may travel more gently over the ground. The degree of abrasion on the sea bed can also be controlled to some extent by the slackness of the lower wings and by adjustment of the lower bridle chains. Tickler chains are often used to dig up the sand just ahead of the ground rope. This drives flatfish and some shrimps and scallops up into the mouth of the net. Vessels working for Dover soles may use several chains, each slightly shorter than the other and thus dragging a foot or so ahead of each other in front of the net. Worn chains or chains of softer steel or alloy are not so effective.

Beam trawls fitted with a series of tickler chains are extremely effective in capturing sole which other trawl nets might scarcely touch. This may not be a bad thing in itself as the sole which bury themselves in the sand might not otherwise be harvested in quantity. What may merit concern is the possible damage to other marine life on the sea bed, from the disturbance caused by this heavy gear.

PURSE SEINES AND RING NETS

Purse seines and ring nets or lamparas are selective both by their use and their design. Most large purse seiners operate chiefly on one species like tuna, anchovy or salmon, or on two or three species like herring, mackerel, pilchard and sprat. With purse seines, one is dealing with schooling fish of one species or with a body of mixed fish attracted by light or FAD. Most purse seine skippers know what species of fish they are setting on, at least 90 percent of the time, from the nature of the echo-traces, the locality, time of year, and the behaviour of the school. Exceptions to this are the smaller light-fishing purse seiners in the tropics or sub-tropics. They take a very mixed catch of sardines, anchovy, horse mackerel, bait fish and some larger species like tuna and mackerel. Occasionally a lot of very young tuna are taken.

It has sometimes happened that a North Atlantic seiner, searching for herring or mackerel, has netted a school of whiting or cod. But this occurs so rarely and in such unusual circumstances that it does not present a management problem. Mesh sizes are important to a seiner skipper mainly because he wants to avoid gilling fish whatever happens. A mesh that is too small poses a very minor problem compared with one that is not small enough. If a catch of very small fish is taken (and schools tend to be made up of fish of one size) then it is a simple matter to let them go while they are still alive. There is much more likelihood of fish surviving if released from a purse seine than from a trawl net after it has been hauled up. Both the density of fish in the bag, and the sudden pressure decrease from the seabed to the surface, can cause mortality before the cod end is released.

Purse seines in Lake Kinneret carry a filter panel at the top of the bunt to permit the escape of small fish. The filter panel mesh is 20 to 25 percent larger than the other seine meshes. Small fish are allowed to swim through the panel before the catch is dried up in the bunt. Some fish are gilled but they can be shaken out of the filter netting. This could hardly be done if the larger mesh was used throughout the body of the net.

An interesting example of how a purse seine can be made to be species-selective is seen in the United States tuna fishery. Schools of dolphin or porpoise were sometimes netted along with tuna and there was considerable pressure from conservationist and environmentalist groups to prevent unnecessary slaughter of these animals. The solution adopted involved a "backing" operation and the use of purse seines fitted with shallow escape lips for the porpoise schools. With skill and practice it became possible to release the dolphins alive without losing any of the valuable tuna catch.

SELECTIVITY OF GILL-NETS

Gillnets are obviously size selective and will seldom capture fish small enough to swim through the meshes. Fish too large to be gilled properly will sometimes be entangled in the netting but this is not always so.

In the early days of the Lake Kariba fishery, some of the indigenous Zambesi fishes grew to remarkable sizes. Tilapia mossambica was one of these and being highly marketable it was much sought after. In 1961 the gillnets used were of 4 and 5 inch mesh chiefly as these were the sizes used in Lakes Mweru and Bwangwelu. During 1962 and 1963 nets of 5½ and 6 inch mesh were successfully introduced and these caught large tilapia which the other nets rarely took. Nets of 6½ and 7 inch mesh were then tried out and even larger fish taken, but their numbers were much smaller. Experimental nets of 7½ and 8 inch mesh were also tested but these caught no fish at all. The experience is mentioned to indicate how large fish can avoid capture by a gillnet. Fishermen can get over this to some extent by hanging their nets very loosely.

Another rarely mentioned aspect of gillnet selectivity is that inactive fish are seldom caught. Just as a long line will only hook hungry, feeding fish, a gillnet will only capture fish that attempt to swim through it. During cold weather or in the dry season, fish in most African lakes become very inactive. Their metabolism slows down to the point where they hardly move or eat. During such periods they will rarely be caught by a gillnet, unless they are driven in as happens in some fisheries. Fishermen in northern Newfoundland and Labrador have observed cod behaving in this way in small bays or creeks, at certain times of the year.

In temperate waters, gillnets tend to be used to capture a single species - herring, cod, salmon or plaice. In the tropics a multitude of species are caught. Thus a mesh restriction to protect juveniles works well in temperate fisheries or in a single species fisheries but is rather pointless elsewhere.

Since the decline of drift netting for herring, the main gillnet fisheries in temperate waters, have been those for salmon and for cod. The bulk of gillnet fishing in the world today occurs among the small-scale fisheries of Asia and Africa. There, such a wide range of species and sizes is taken that control by mesh restriction is out of the question in all but a very few isolated cases.

SELECTIVITY BACKGROUND INFORMATION

Selectivity is one of the main components of a stock assessment model related to the fishing process and different assumptions can have a large influence on the stock assessment results and management advice. Selectivity is used to model the vulnerability of fish to the gear as well as the availability. The availability may be related to the spatial stratification of size or age of fish and is modeled by having separate fisheries for each area with their own selectivity curves. Selectivity assumptions range from the freely estimated and time varying assumptions of a Virtual Population Analysis through nonparametic approaches to functional forms. Selectivity can be a function of length, age, or both. Selectivity may also differ between the sexes, change over time or space.

SELECTIVITY MODELS

There is a large range of selectivity models that have been used in fisheries stock assessment. Virtual Population Analysis essentially treats each age as a separate parameter for each year. At the other extreme, simple age-structured models assumed selectivity was knife edged at a fixed age (i.e. all fish less than a certain age are not vulnerable to the fishery and all fish that age and older are fully vulnerable to the fishery). As statistical catch-at-age stock assessment models started to be developed, more attention was applied to the form of the selectivity curve and the estimation of the selectivity curve's parameters within the stock assessment model. The logistic curve, which becomes asymptotic at older ages, became popular because it represented the data obtained from trawl selectivity experiments. However, not all fishing gears (e.g. gill nets) efficiently catch the oldest individuals and behavior or spatial distribution may make the old individuals unavailable to the gear. Therefore, dome shape selectivity curves were developed. These include the double normal (Coleraine), which is a normal distribution with different variances for the left and right hand sides, and the exponential logistic (Thompson XXXX).

Non-parametric approaches have also bee used to model selectivity. For example, several assessments have estimated each age as a parameter, while applying smoothness penalties to constrain how selectivity changes from age to age. MULTIFAN-CL uses cubic splines to model selectivity.

ASYMPTOTIC SELECTIVITY

Historically, selectivity was often assumed to be asymptotic. However, as stock assessment models started to become more complex and model different gear types and areas as different fisheries, it become apparent that many of the fisheries had dome shape selectivity curves. However, if all fisheries are dome shape, the descending limbs of the selectivity curves are confounded with natural mortality (Thompson XXXX), catchability, and other model parameters. It is therefore often considered necessary to assume that one of the fisheries has an asymptotic selectivity to stabilize the parameter estimation.

It is also considered a conservative assumption with respect to over exploiting the stock. Sometimes it is difficult to determine which fishery should have asymptotic selectivity and incorrectly assuming a fishery's selectivity is dome shape can substantially bias the results (Wang et al. 2009). This may be the case even if that fishery does not capture a large proportion of the catch. The right hand limb of the selectivity curve determines how many large/old fish are captured and influences the fit to age/size composition data. Using an asymptotic selectivity when the actual selectivity is dome shaped will tend to cause estimates of fishing mortality to be biased high. In some fisheries, tagging data suggests that selectivity for all fisheries is dome shape. However, this is confounded with increasing natural mortality with age.

TEMPORAL VARIABILITY

Selectivity may change over time due to modifications to gear, changes in the season or area fished, temporal variability in growth, or other factors. Therefore, modeling these changes may be important in the stock assessment.

In general, fisheries are defined so that their selectivity does not change over time. For example, if a fisheries selectivity does change for some reason (e.g. mesh size changes) the fishery is often divided into two fisheries with different selectivity curves estimated. However, it may be desirable to share information among the two periods (e.g. selectivity for the older fish remains the same) and therefore maintaining a single fishery and allowing the selectivity curve to change may be desirable. One or more of the selectivity curve's parameters can be allowed to vary over time. Non-parametric approaches similar to those used for modeling selectivity as a function of age can be used to model changes in selectivity over time.

Virtual Population Analysis (VPA) implicitly assumes that selectivity changes each year. However, this requires that catch-at-age data is available for every year. VPA also combines all catch into a single fishery even if the fisheries catch different sized fish. Allowing selectivity to change over time reduces the information content of the data. Therefore, there is a tradeoff between maintaining information and biased results due to real changes in selectivity. The reliability of age composition data may be low for some fisheries and there is a tradeoff between fitting the composition data to estimate the selectivity curve and remove the fish at the right size and bias caused by allowing the composition data to influencing the estimates of recruitment and exploitation rate. If the catch from this fishery is only a small component of the total catch, removing the catch at the right age may not be important.

USING SELECTIVITY TO REPRESENT SPATIAL VARIABILITY IN AGE AND SIZE

The size or age distribution of a population often changes across different areas. For example, juveniles and adults often reside in different areas, perhaps to avoid cannibalism. Therefore, even though fisheries in different locations are fishing on the same stock, they catch different aged fish and their effective selectivity differs. In this case, it is common to separate the areas into different fisheries and estimate different selectivities for each fishery. this is particularly important if the relative fishing mortality rate among areas chnages over time.

DISCARDS

Discarding of fish can contribute substantially to the mortality in some stocks and should be taken into consideration in stock assessments. The size of discarded fish is a function of the selectivity of the gear and the sizes of the fish that are retained. The size of fish retained may be driven by regulations on minimum legal size or on market preferences. In some cases, discards may be due to sorting of fish by size as well as fish that are not marketable (e.g. spoiled due to inability to process the fish quickly enough). Different selectivity curves would apply to these two types of discards. The selectivity of discards is often difficult to determine since they are not landed for sampling and special onboard sampling programs may be needed.

LENGTH BASED VERSUS AGE BASED SELECTIVITY

Selectivity can be modeled as a function of length, age, or both. Selectivity of nets generally has a length based component. Small fish go through the mesh and large fish may be too large to be meshed (e.g. in set nets). In other cases, fish may migrate once they reach a certain age and selectivity for fisheries in different areas is age based. In most cases it is difficult to know which process is more important, age or length. When length is a fixed function of age, it may not matter which is used. However, if growth or the variation of length-at-age is being estimated, it may be important to use the correct basis for modeling selectivity.

Spatial differences in the ages of fish may be due to age specific movement while vulnerability to the gear may be length based. In this case, a composite selectivity curve composed of an age based selectivity curve and a length based selectivity curve may be appropriate.

ESTIMATION

The estimation of selectivity curve parameters is the same as other model parameters, but they often cause convergence problems. Flexible functional forms for selectivity curves often fit the data better based on statistical tests (e.g. likelihood ratio and AIC), but they often have sharp changes and cause the likelihood surface to be rough, which causes multiple local minimum. This is a frequent problem that needs to be resolved.

The nonparametic methods used to model selectivity have "smoothness parameters" that control how much the selectivity can vary. These parameters are usually arbitrarily set to provide selectivity parameters that look reasonable. Maunder and Watters (XXXX) used cross validation (based on length-frequency data) to determine the smoothness parameters.

MANAGEMENT IMPLICATIONS

Bias in the stock assessment results caused by misspecifying the selective curve can directly bias the management advice. Appropriately modeling the selectivity for different components of the fishery is also important because management may direct regulations at different components of the fishery.

SELECTIVITY OF FISH ATTRACTION METHODS

The three main methods of fish attraction are by light, by bait, and by floating rafts and artificial reefs. Each method is selective under different conditions. The selectivity is the product of the combined effect of the attractant, the fishing gear, the location and the method or application. Note also that secondary attraction occurs. Larger species come to prey on small fish congregating around the attractant.

LIGHT ATTRACTION

Lights are used throughout the tropics and sub-tropics to attract fish for capture by purse seine, ring net or lift net. Light attraction is not much used in temperate waters owing to negative or unpredictable reaction by fish. In many cases there juveniles are attracted and mature fish repelled or left unaffected. Most of the small pelagics in the tropics display positive phototaxis. These include sardinella, anchovy, saury, small mackerel and bait fish. Squid are also attracted to light but prefer to stay in the fringes of the illuminated area. Larger fish are attracted by the concentration of small fish, and by their movement or increased activity in the presence of light.

Thus catches made with light attraction and seines or lift nets tend to be of mixed species of fish. Exceptions to this are squid which are taken by special jigs that would not catch other species, and surface swimming saury which are taken by stick-held dip nets in waters north of Japan. Catch composition may differ from area to area. The same net and lights may catch different fish if employed over shallower or deeper waters.

In Lake Tanganyika, dagaa and ndakala sardines are taken along with perch and Nile perch. The dagaa of Lake Kariba are accompanied often in light-attracted catches by quantities of Alestesspecies. In S.E. Asia, sardinella, mackerel, scad, anchovy and tunas are taken together. Light-induced catches in the Mediterranean may consist of mackerel, pilchard, sardinella, anchovy, sprat, immature barracuda and occasional tuna.

Small light-sticks are now used on certain baited long lines, drop lines and traps. They are used on swordfish long lines and on hand lines and fish traps for large demersal species in places like the Caribbean.

BAIT ATTRACTION

Bait is possibly the oldest means of attracting fish and it is still in widespread use today, particularly in traps and pots used for lobster, crabs, prawns and rockfishes. Longlining for cod, halibut, ling or dogfish in northern waters is coming back into use after a decline of some years. Tuna long lining is still practiced in the Pacific and Indian Oceans. Swordfish long lines are used off the eastern seabord of the USA. Small-scale fishermen in some parts of the world use bottom-set long lines.

Most of these methods are fairly species selective. While one still gets crabs in lobster pots, skate on cod lines, and shark on tuna lines, they are the exception rather than the rule. The baited lines and traps can be considered to be selective in both attraction and capture. This is partly the case, of course, because they are set in areas and at depths where the fish sought are known to be present. Live bait fish are used to attract schools of skipjack, bonito and other tuna for capture by pole and line. This is a very selective method, since the schools are detected visually first and since the fishermen can see what they are taking.

Some of the live bait used is itself caught by bait - a ground meal or powder made of cereal, fish or other protein. The use of this kind of bait is usually called "chumming" and it is normally selective in attracting only the small bait fishes needed by the pole and line vessels.

FADS AND ARTIFICIAL REEFS

Fish aggregation devices, or FADs for short, are used to attract pelagic fishes in the tropics. Tuna purse seiners and pole-and-line vessels fish around those anchored in deep waters off the islands of the Indo-Pacific. Smaller purse seiners and lift-net boats fish around FADs anchored in shallower waters nearer land where sardines and small mackerels may be found. The FAD itself is a floating raft to which are attached strings of palm leaves or old ropes, netting or tyres. The attraction of fish to these devices is not fully understood. It may be partly for shelter or protection, partly for orientation, and partly to feed on the smaller fish or algae that accumulate there.

As with light attraction, a mixed body of fish is attracted to a FAD. Light or bait attraction is often used to supplement the drawing power of the FAD, or to draw fish in closer before capture. The small fish usually swim closest to the raft, and the larger ones further away. Big tuna may be caught by hand line many fathoms below a FAD.

Artificial reefs, as their name implies, attract mainly reef fish and crustaceans. They are installed on bare stretches of the sea bed where the natural habitat is poor. The artificial structure becomes a favourable habitat for many of these species. Although few of them have been constructed in northern Europe, fishermen there for many years have practised fishing close to rocks or wrecks where fish tend to congregate.

The selectiveness of fishing methods used in conjunction with attractants may be summarized as follows:

Very selective	: Squid fishing with jigs and lamps Stick-held dip net and lights for saury Live-bait pole and line fishing for skipjack Chumming for small bait fish
Generally selective	: Baited and/or lighted long lines Baited traps and pots Baited, trolling lines and hand lines
Less selective	: Purse seining or lift-net fishing with the use of light attractionPurse seining around FADs Gill netting and hand lining around reefs and wrecks

EFFICIENCY AND SELECTIVITY OF A TRAP AND TRUCK FISH PASSAGE SYSTEM IN BRAZIL

In Brazil, the main emphasis in the management of reservoir fisheries seems to be on the rehabilitation of the fish species affected by the formation of dams. The companies are obliged to protect threatened fish species either by incorporating fish ladders in the dam design for facilitating easy migration, or by creating breeding facilities to produce fingerlings of the affected native species to be stocked in the reservoir (Sugunan, 1997).

Although only a small fraction of South American species are migratory (Petrere Jr., 1985; Godinho & Godinho, 1994), they are the most important for professional as well as amateur fishing, due to their larger size and abundance (Northcote, 1978).

For decades, the environmental legislation made fish stocking and the control of fisheries the main, and in some cases, the only strategy for the conservation of these species (Agostinho et al., 2007). Although these policies continue to be pursued throughout the country, studies are lacking that determine their efficiency in the recuperation of species and threatened populations, as well as their cost–benefit ratio and their role in the sustained management systems of fish populations (Vieira & Pompeu, 2001). The low fishery yield in reservoirs located in the South and Southeast regions of Brazil, with a significant reduction of migratory species (CESP, 1996; Agostinho et al., 1994) clearly indicates that this strategy is not satisfactory.

The construction of fish passages was another strategy adopted by the Brazilian energy sector in order to diminish the effects of barriers on fish communities, mainly the migratory species. The first fish ladder in Brazil was constructed in 1911, at the Itaipava Dam, in the Pardo River, upper Paraná Basin (Godoy, 1985). In 1927, the construction of fish ladders became a legal requirement in São Paulo State (Agostinho et al., 2002). With the increasing number of hydroelectric facilities in the 1960s, the necessity of fish passages was incorporated by the legislation of other states. Currently, several such structures have been installed in Brazil, including ladders and lifts.

However, evaluation studies are scarce, and as in other actions connected to the management of fishing resources in reservoirs, most of the fish passages were never monitored, despite the high investments and efforts involved (Agostinho et al., 2002). In this context, efficiency can be defined as the proportion of fish that utilize the fish passage facility (Novak et al., 2003), and selectivity refers to the selection of different species or size classes.

The studies performed in South America have mainly concentrated on fish ladders, focusing on their selectivity concerning a particular fish species. In this paper, we evaluated the efficiency and selectivity of the first trap and truck fish passage system in Brazil, installed in the Santa Clara Power Plant, Mucuri River.

MATERIAL AND METHODS

Study Area

The Mucuri River is part of independent basins which drain the east region of Brazil, having a total drainage area of 15,100 km^2 (CETEC, 1983). The basin's hydrological regime is characterized by two well–defined seasons, dry and rainy, with larger flows from November to April, when the breeding of most of the region's fishes takes place (Pompeu, 2005).

The river can be divided into two environmental units, according to their physiographic characteristics: its upper and middle courses show a larger declivity, rapids and a predominantly rocky bottom; its lower course, however, is characterized as a deposition system, with small declivity, predominantly sandy bottom, with the existence of a seasonally flooded Atlantic Rainforest (Mata Atlântica), utilized for cocoa cultivation, and wetlands and mangroves near the coast.

Santa Clara Dam is located on the Mucuri River, in a transitional zone between these two physiographic units, approximately 80 km from the river's mouth on the Atlantic Ocean. Dam construction began in 1999, and the first turbine became operational in February, 2002. The concrete dam has a maximum height of 60 m and is 240 m long. In November of 2003, a fish passage with a trap and truck system became operational in the Santa Clara Hydroelectric Plant.

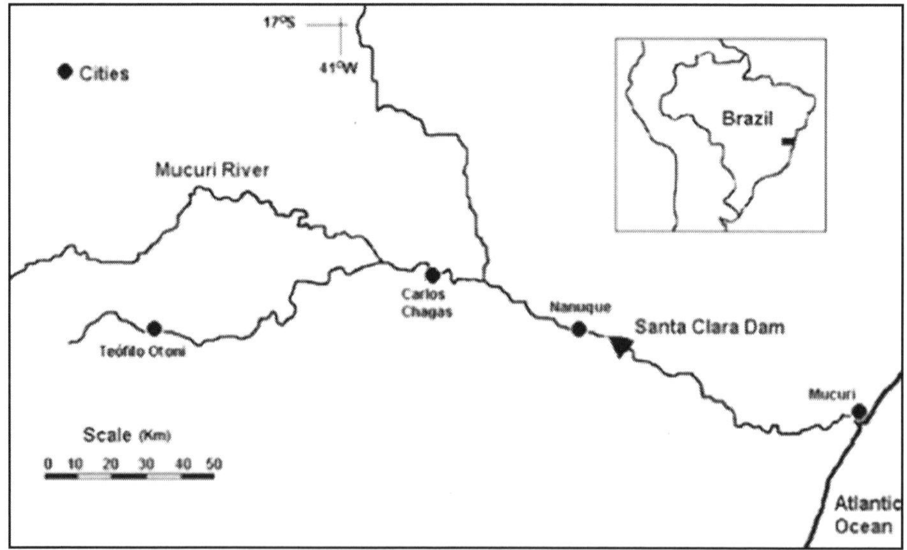

Fig. Mucuri Basin, Indicating the Location of Santa Clare Dam.

It is located 200 m downstream of the dam, immediately after the tailrace. It includes an attraction (collection) channel, an elevator, and a truck equipped with a tank. An operational cycle of the fish passage takes about

35 minutes, and consists of the following sequence of events. Fish are attracted to the entrance of the collection channel by a 3 m^3/s flow, supplied from the reservoir.

The collection channel is 1.8 m wide, 16 m long, and its depth varies with the Mucuri River level, from 40 cm up to 3.5 m. A fish elevator is located in the upstream extremity of the channel. The operational cycle begins with the movement of a mechanical screen, located 8 m upstream of the channel entrance, which crowds the fish in the direction of the elevator. The elevator has a capacity of 4 m^3, and requires 10 minutes to travel from the collection channel to the level where the fish are transferred to a truck equipped with a tank with the same capacity.

When the elevator is empty, it is returned to its initial position, and the truck carries the fish to the reservoir, in an approximately 15–minute trip, where they are released 500 m upstream of the dam. It has been operated during the rainy season, with cycles every two hours during the day.

Sampling and Analyses

The species composition in the Santa Clara Dam's fish lift was compared to the original composition of lower Mucuri River fish fauna and with the populations that gather downstream of the dam, in the tailrace, during the reproductive season (October to March).

Fish samplings were performed every three months, between October of 1998 and July of 2001, in the lower Mucuri River at three stations, located 1, 40 and 80 km downstream of Santa Clara Dam, respectively. Fish were caught with gillnets (20 m long, with 3 to 12 cm stretch mesh), seines (5 m long and 1 mm mesh), cast nets (3 cm stretch mesh) and hand nets (1 mm mesh). Gillnets were fished in the water column for 14 h overnight.

Seines were used in shallow areas or littoral zones, hand nets were employed in near–shore aquatic macrophytes and in riffles, and cast nets were used in habitats too deep to wade. The three latter methods were employed for 1–3 h and used only qualitatively. After being captured, the fish were immediately fixed in 10% formaldehyde solution and subsequently preserved in 70% ethanol. In the laboratory, each fish was measured for standard length and identified. Catch per unit of effort (CPUE) in number (CPUEn) was used to express data on numerical abundance (number of individuals per 10 m^2 of net in 14 h; CPUEn).

For the evaluation of the fish community which concentrated in Santa Clara's tailrace, between October 15th of 2002 and February 15th of 2003, fish were caught by four professional fishers using cast nets with mesh sizes of 6 and 8 cm (between opposite nodes). The fishers caught the largest possible number of individuals, in a pre–determined 300–m course immediately downstream of the dam, including the tailrace. These fishers began at 7:30 a.m.

and finished at 5p.m., during the entire period, from Monday to Saturday. After the capture, the fish were counted, identified, measured and transported to the reservoir, where they were released to avoid harmful effects.

For the evaluation of the fish lift's efficiency, immediately before the beginning of its operation, between October 16th and November 7th of 2003, fish were captured downstream of the Santa Clara Dam with cast nets (3 cm stretch mesh). All fish were tagged with an external hydrostatic tag and released in the river at the same place where they were captured. The proportion of these tagged individuals transported by the lift was used to estimate its efficiency by species, for the group of migratory ones and total fish fauna of the Mucuri River.

On November 19th of 2003, the operation of the Santa Clara Dam fish lift was initiated, and conducted for 4 months, until March 19th of 2004. During the operational period, from Monday to Saturday, six daily passing cycles were carried out, at 8 a.m., 10 a.m., 12 p.m., 2 p.m., 4 p.m. and 6 p.m. At 15–day intervals, night passage cycles were also performed, at 8 p.m., 10 p.m., 12 a.m., 2 a.m., 4 a.m. and 6 a.m. A total of 636 passage cycles were performed, 588 day and 48 night cycles. At each cycle, before releasing of the fish in the reservoir, they were counted, identified and measured. The dead or injured individuals and the tagged ones were recorded. Regression analysis was used to test the relationship between the number of dead or injured individuals and the number of fish transported in the truck's tank.

Although the movements of the Mucuri fishes were not studied, they were considered migratory, and the lift's target species, the native ones from the genera Leporinus, Prochilodus and Brycon, have representatives with known migratory behavior (Lucas & Baras, 2001). For four of these species (L. conirostris, L. steindachneri, P. vimboides and B. ferox),the selectivity in size was evaluated by comparing the standard size classes of the individuals observed in the lift with the size classes of the individuals caught in the Mucuri River between 1998 and 2001, and with the fishes caught with cast nets immediately downstream of the dam in 2002/2003. The lift's selectivity was also evaluated by comparing the richness of the fish assemblages caught in the lower Mucuri River and in the tailrace, with the one passed by the mechanism.

RESULTS

In the lower Mucuri River, 49 fish species were recorded, including 34 native freshwater, five exotic and ten marine.

Of the Mucuri's fish species, 33 were captured in Santa Clara's tailrace (2002/2003 reproductive season) and 32 were passed by the fish lift (2003/2004 reproductive season). During both occasions, only six species were responsible for more than 95% of the individuals, a pattern different from the lower Mucuri River's assemblage. However, while four migratory species (P. vimboides, L.

conirostris, B. ferox and S. steindachneri) were found in great abundance during the two events, Pogonopoma wertheimeri was considerably more abundant in the captures conducted in the tailrace, and the lambaris (Astyanax spp.) were only recorded in the fish lift.

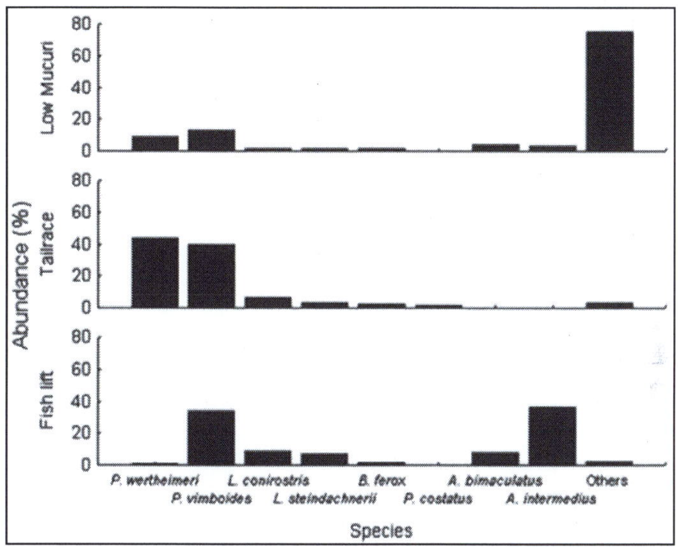

Fig. Abundance (%) of captured species in the Santa Clara tailrace (2002/2003) and passed by the fish lift (2003/2004).

The species recorded in the fish lift represented 66% of the region's native fish richness. Species which were not recorded in the lift were, in general, of smaller sizes. Besides, only five out of ten marine species present were passed.

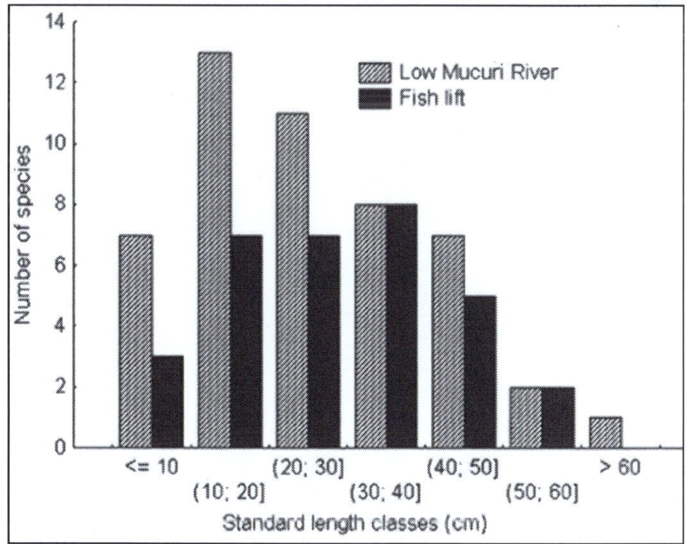

Fig. Number of species by size class found in the lower mucuri River and passed by the Santa Clare Dam fish lift.

Freshwater migratory species, the main target of the fish passage, comprised approximately 50% of individuals captured in the tailrace with cast nets (2002/2003), the same proportion found in the fish lift (2003/2004), while in the river, they represented only 19% of captures with gillnets. For the migratory species, the relative abundance per size class was also similar when compared to the fish caught in the tail race and passed by the fish lift. However, in comparing these fish assemblages with the natural composition of the river, less representative captures of smaller–sized individuals were observed.

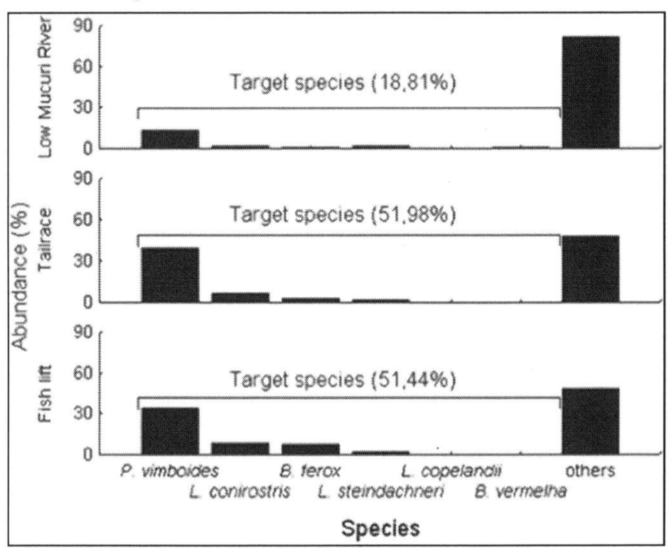

Fig. Relative abundance of targer species in the lower mucuri River, in the Santa Clara tailrace (2002/2003) and in the Santa Clara Dam fish lift (2003/2004).

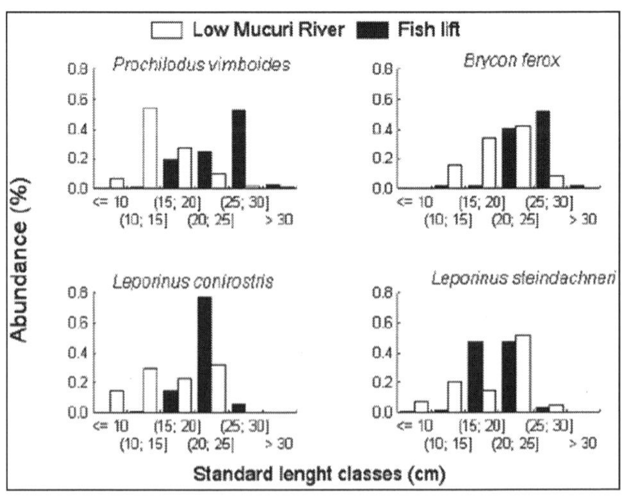

Fig. Reletive abudance of migratory species per size class for the lower Mucuri River, in the Santa Clara tailrace (2002/2003) and in the Santa Clara Dam Fish lift (2003/2004).

No significant relationship was seen between the number of fish passed per cycle and the mortality rates during their capture by the fish lift and transportation ($F = 1.898$; $p = 0.169$). Of the total of passed individuals, only 0.48% died or were injured during the complete fish passage cycles. However, marine species and the Siluriforms suffered more injuries, when compared with the migratory Characiforms and the total passed community.

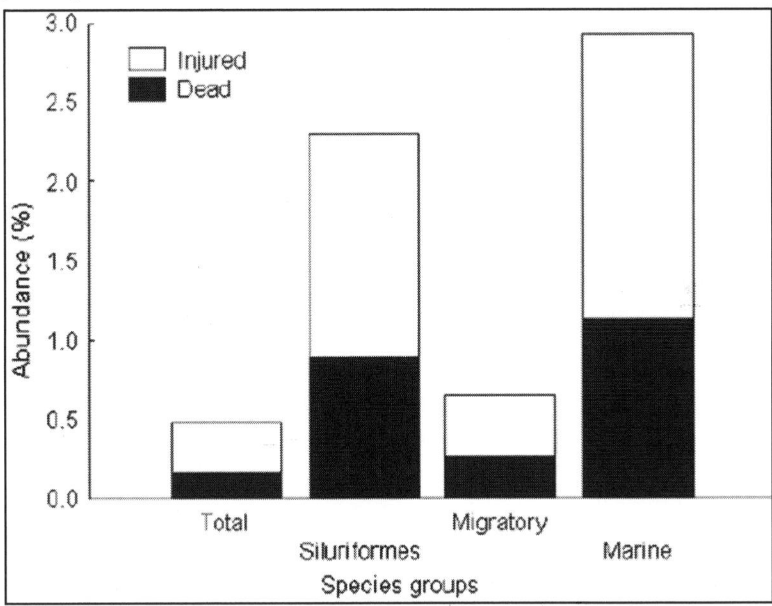

Fig. Nimber of injured of dead fish after being passed by the fish lift according to species groups.

During the period prior to the lift's operations, 1718 individuals of 21 species were tagged. However, only five species had individuals captured during the passage. The efficiency of the mechanism, estimated through the capture rates, varied from 0.2% for the cascudo–preto (P. wertheimeri) up to 16.1% for the piau–branco (L. conirostris), with 7% for the group of migratory species and 3.1% for the region's fish community group.

DISCUSSION

In a general way, the fish lift can be considered to have a low selectivity in relation to species, but it also showed difficulty in capturing small–sized individuals. The mesh size range used in gillnets (3–12 cm) for the lower Mucuri River, can capture smaller fish when compared to cast nets (6–8 cm) used in the tail race. This way, the similarity between the assemblages passed by the lift, and sampled in the tail race must be reflecting comparable selectivity. The efficiency of a fish lift concerning the capture of small–sized individuals is related directly to the dimensions of the screen mesh, which is supposed to be sufficiently small to prevent fish passing through it. This can lead to difficulties since very small fish require very small screen mesh sizes, which would result in maintenance difficulties (Larinier, 2002). It is important to point out that the fish lift was highly selective for Pogonopoma wertheimeri, a non migratory bottom catfish, which was very abundant downstream of the dam, even in the captures with cast nets, but showed a very low recapture rate by the lift. A few studies have focused on the efficiency of fish passages in South America, in terms of number of fish passed and their capacity to utilize the mechanism. These studies have also indicated that many individuals of a few species have overcome the passage and a variable selectivity regarding species composition.

In the case of fish lifts, the mortality rates must also be taken into account when evaluating the mechanism. In the case of the Santa Clara fish lift, marine species and Siluriforms were particularly more sensitive, although the mortality rate for the group of passed species was kept extremely low. Even the density in the elevator and in the truck's tank does not seem to have influenced the survival capacity of the individuals during the passage process. Considering the fact that the Siluriforms, possibly because they swim close to the walls (Santos et al.,in press), have a greater probability to suffer damage due to the motion of the mechanical parts, the absence of migratory fish in this order must have contributed to minimize mortality rates or fish injuries.

Quantitative estimates of the efficiency of fish passages are only possible through tagging and recapture programs or when there is another fish passage downstream, where the number of passed individuals has been determined. Even in North America, these kinds of studies are scarce. From a databank of 213 projects for which at least one paper concerning fish passage has been published, information on the number of fish which utilized the mechanism was only available for eight mechanisms, where only three provided sufficient data for a quantitative estimate of the efficiency (Novak et al., 2003).

When compared to the little information available in South America, it is possible to state that the Santa Clara mechanism seems to be more efficient than the only system that has been evaluated with similar methods. According to Oldani & Baigún (2002), only 0.68% of migratory individuals and 1.88% of individuals belonging to all species of the La Plata River, utilize the two fish lifts implemented in the Yaciretá Dam, in Argentina. Species considered target species, despite representing 30% of the captures with nets, downstream of the dam, comprised only 10% of the total number of fish passed.

Some causes were pointed out for the low efficiency of the Yaciretá fish lift:

(a) "Fishway dimensions are undersized by a large margin for the number and size of fish that should be passed. Elevator cross-sectional areas are small compared to the attraction channel dimensions with the left elevator having only about 1.8% of the approach channel cross-section and the right elevator only 1.1%". In the case of Santa Clara, the lift channel's section represents between 4 to 5% of the river's section, and the fish community does not include large–sized species. However, more relevant is the difference in water volume available for attraction. While in Yaciretá, 1 to 2 m^3/s are used for attraction in each lift, which represents an overall value less than 0.04% of the average annual flow of the Paraná River, in Santa Clara the attraction (3 m^3/s) represents 2.2% of the average flows in the Mucuri.

(b) "Number of fish transferred by the system is limited by the number of cycles that can be completed. Even in optimal conditions, elevators cannot perform more than 24 cycles per day". In the Santa Clara lift, the same limitation is present. However, while the truck transports fish to the reservoir, the elevator can return to its initial position and start a new fish attraction process.

(c) "There is no way to regulate the number of fish that enter the elevator. Overcrowding of fish in elevators can cause serious stress and injury..." In the Santa Clara lift, during periods of large flows of fish, the mechanical screen can be lowered without running the main confining process, avoiding the entrance of new fish and diminishing this problem.

(d) "Some species of fish that are strongly rheotactic are reluctant to leave the elevators and swim through the exit channel into the upstream reservoir". In the Santa Clara fish lift, the fish are transferred from the elevator's container to the truck's tank and further to the reservoir, avoiding this problem. However, some fish, due to rheotaxis, remain in the truck's tank, which are then manually transferred to the reservoir.

(e) "The elevator entrances are very close to the turbine draft tubes so that the influence of attracting flows exiting the elevator entrance may be masked by powerhouse discharge". In the Santa Clara fish lift, this situation does not occur since the elevator's entrance is situated about 200 m from the powerhouse outtake. However, throughout the studies, large shoals could be observed next to the draft tubes, raising doubts concerning the best region for its placement, because it is not clear if they are able to move back to the region around the lift.

(f) "The geometry of the entrance to the collection channel may limit or prohibit entry of thalweg–oriented fish such as large siluriforms, since the fishway channel entrance does not connect to the river–channel bottom." The Santa Clara lift's connection channel entrance seems to perform the same limiting effect for the bottom fish's ascent, which seems to have been particularly important for the cascudo–preto P. wertheimeri.

The relatively complex operation of fish lifts, including the innumerous mechanical and electrical systems, results in frequent interruptions of the passage activities, which can last for long periods of time. During the reproductive period studied (four months), in ten days, there were at least 2–h interruptions in the passage activities. Its maintenance costs are also quite high when compared to fish ladders. This is the reason why static mechanisms (without movable parts) have been more utilized in France, where although they are more expensive in terms of engineering, they are more easily maintained and reliable (Larinier, 2002).

On the other hand, when compared to fish ladders, the fish lifts have the advantage of the possibility of adjusting the number and time of passage cycles according to the largest fish flows, a procedure which can, in some cases, mean the saving of large amounts of attraction water, which can be used for generation purposes (Pompeu & Martinez, 2005). The trap and truck system can also be considered a good option for passing fish in hydroelectric plants with a powerhouse distant from the dam. When the probability of spills is low even in periods of flooding, which is a common situation in small hydroelectric plants, the efficiency of a fish passage, installed next to the dam, is compromised, since the fish frequently have difficulties in reaching the mechanism. In these cases, the fish passage with a trap and truck system allow the attraction and capture of fish next to the powerhouse, where shoal agglomerations are usually observed, and their transportation to the reservoir (Pompeu & Martinez, 2003)

Another important aspect of fish lifts with a trap and truck system is the possibility of controlling which and how much fish will be passed. For migratory species, depending on the dam's location, it is essential that the obstacles are overcome in such a way that the spawning sites are reached in the desired time. However, for the resident species (which perform more local trophic or reproductive movements), the main biological goal must be to avoid population fragmentation. In this case, a fish passage system can be considered efficient if it is utilized by a certain number of individuals and not necessarily by the entire population (Larinier, 1998), and the passage must be controlled to avoid fishery depletion. Considering the difficulties in promoting downstream migration through the Santa Clara dam, which occurs only during large spills (Pompeu et al., 2004), the passage of marine species upriver should be strongly limited, because their breeding sites are located exclusively downstream of the dam, in the estuary. Therefore, the fish lift management plan should consider the passage in large number of only the target species, for fish and conservational reasons.

As important as the fish's capability to locate the entrance of the mechanism and pass the dam is the evaluation of the importance and effectiveness of the mechanism in maintaining the populations of migratory species, an aspect rarely evaluated. For the migratory species of the Mucuri River, the spawning sites are located upstream, and the nursery areas downstream of Santa Clara Dam. However, the passage of eggs and larvae through the reservoir and dam's spillway seems to be possible (Pompeu, 2005). Considering the selectivity and efficiency values of the fish lift in attracting shoals, this mechanism will probably contribute to the conservation of the migratory fish species of the Mucuri River. However, only a long–term monitoring of the ichthyofauna will be capable of revealing its efficiency in maintaining viable populations.

Fish passages with a trap and truck system have been proposed for a great number of hydroelectric power plants. In Minas Gerais State, they were indicated for at least ten facilities in licensing or implementation processes in the Doce, Paraíba do Sul and Jequitinhonha River Basins. Because they belong to the same bio–geographic domain, called the East Basins, these basins share the majority of migratory species with the Mucuri River (Bizerril, 1994). Therefore, it is expected that the information obtained in this study will be of great importance for the evaluation of the convenience, as well as providing subsides for the implementation and operation of this kind of mechanism in other hydroelectric plants.

Additionally, this information would also be quite useful for evaluating the efficiency and better planning of conventional lifts, which have also been implemented in various dams, and which have identical attraction and capture systems.

EFFECTIVENESS OF MESH REGULATIONS

Minimum mesh sizes have been introduced in most well-developed trawl fisheries to reduce the fishing mortality of undersized or juvenile fishes. That small fish can escape through cod-end and bag meshes has been confirmed by numerous underwater films and observations. Whether the mesh regulation has a significant impact on the number of juveniles surviving to maturity is not always certain. Other factors must be examined, and these include the post-release mortality rate, the percentage of juveniles that would be caught in trawls, vis-a-vis the total juvenile population, and periodic natural fluctuations in the stock of fish in question. An important factor in fisheries where adjoining countries harvest a common stock of fish, is the existence and enforcement of similar legislation in each country concerned. There is little incentive for one state to protect young fish if its neighbour is harvesting them regardless.

As trawl nets differ greatly in performance, we will examine the effects of minimum mesh sizes on three main types of bottom trawl: multi-species bottom trawls; bottom trawls for pelagic species; and shrimp trawls. Note that in each case the minimum mesh size applies to the bag and cod end only. The meshes in the wings, square and shoulders are much larger.

BOTTOM TRAWLS FOR DEMERSAL SPECIES

These trawls are used all over the world, chiefly by inshore or nearwater vessels, but the variety of species caught tends to increase towards warmer waters. A further complication in the tropics is that very small fish are quite saleable on local markets. In temperate countries consumers tend to avoid small fish, so if these are caught they generally go for reduction to meal, oil or pet food.

In a temperate fishery most fishermen would accept a mesh restriction without much argument. There are two situations in which it might be resented. One concerns fish which owing to their shape tend to escape through meshes which would stop mature fish of a similar species. The whiting, Gadus merlangus, is an example of this. Though it resembles the haddock, Gadus aeglefinus, a mesh that would retain mature haddock would permit fairly large whiting to escape.

Conversely, a mesh that would retain mature whiting might also retain immature haddock. Further legislation introduced to cope with this problem has involved minimum sizes for fish sold for human consumption. However, that does nothing to prevent numbers of small haddock being killed in order to capture mature whiting. In that kind of situation, if the mesh regulation aimed at haddock is obeyed, large quantities of mature whiting will not be caught. If it is infringed, or area by-laws introduced to permit smaller meshes, some immature haddock and other fishes will be destroyed.

The other situation concerns fisheries where the income from industrial fish (fish sold for conversion to meal, oil or pet food) is a significant portion of the gross earnings. This creates an incentive to keep as large a by-catch as possible.

If the by-catch is pout, sand eels or sprat, little harm is done, but if it is young herring, whiting or haddock, then other fisheries could suffer. Legislation to limit the harmful effects of industrial fishing usually relates to the proportion of juveniles of other species in the by-catch. However, unless there is strict inspection, infringements will occur regularly, and this makes enforcement both troublesome and expensive.

In tropical and Mediterranean fisheries, bottom trawlers tend to keep all fish caught for human consumption, regardless of size. This results in the landing of many juveniles and the constant incentive to use small-meshed cod ends. A minimum mesh size regulation is thus difficult to enforce and if enforced it could result in the loss of many mature edible fishes like squid, lizardfish and wolf-herring. Further complications exist owing to the pelagic and semi-pelagic behaviour of many of the species. They can be trapped in seines or lift nets which do not have a mesh restriction. So immature fish escaping from a bottom trawl in the tropics may still be caught in another type of gear before they reach maturity.

BOTTOM TRAWLS FOR PELAGIC SPECIES

These are mostly herring trawls and tend to be Scandinavian vinge trawls or high-opening semi-pelagic nets of the Polish or German type. Since they are used to catch herring, mackerel, sardine, sprat or sand-eels, they must of necessity use small-meshed bags and cod ends. When properly rigged and operating for the prescribed species these trawls take very little other fish of any description. However, it is possible to trim the gear in such a way that it will take demersal fish in quantity. This can happen when there are no herring or sprat around and the skipper wishes to make up a load of any kind of fish. However, as the mesh size remains the same, many immature white fish could be taken. These would either be dumped (dead) at sea or brought in for industrial use. The point is that once trawlers are permitted to use small-meshed bottom trawls for pelagic fish it is almost impossible to prevent their use on demersal stocks if the pelagic species are scarce.

SHRIMP TRAWLS

Three catch problems associated with shrimp trawls concern the by-catch of mature edible fish which is usually discarded, the capture and destruction of immature fish and the capture of juvenile shrimp. The fish by-catch problem cannot be tackled by mesh size limitations, otherwise no shrimp would be caught. Separator shrimp trawls have recently been designed in an attempt to deal with the problem. In these trawls, much of the by-catch is separated from the shrimp and released before it reaches the cod end. But this necessarily involves some loss in shrimp catch, additional gear expense, and the inconvenience of handling and repairing a more complex net.

An alternative solution adopted by some states is to insist that the shrimp trawlers retain all the mature fish, or a significant proportion, and bring them back to port for sale on local fish markets. Only the smallest juvenile fish will escape through a mesh size designed to retain shrimp. Whether the destruction of young fish by shrimp trawls is sufficient to merit concern is debatable. Many young fish are high swimming and shrimp trawls have low headline heights so the juvenile catch may not be excessive, but this could be determined only after proper investigations. Many shrimp fisheries do not have even a minimum mesh size to protect juvenile shrimp. In tropical countries where small shrimp can easily be sold for human consumption and for making shrimp paste, there is not much motivation among fishermen to protect juveniles. Surprisingly, several such fisheries seem to have survived many years of intensive fishing with small-meshed trawls. The Malacca Strait fishery is an example of this.

A mesh restriction for shrimp trawls to protect juvenile shrimp would appear to be in the interests of all concerned, and would have few undesirable side effects. But enforcement would have to be strict throughout the whole area. The fish by-catch problem could not be tackled by mesh regulation, only by use of separator trawls, or by insisting that mature fish be preserved and taken back to port for sale. Unfortunately, both these solutions may be costly or difficult to enforce.

OTHER OBSERVATIONS

Cushing (1974) states that "conservation by mesh regulation is least conservation because it is adapted to the smaller and numerous species like the haddock in the North Sea; larger species (for example, cod or turbot) are not necessarily conserved there as well as they might be."

When considering the effect of a mesh size one must bear in mind not only its dimensions but its shape when the net is being towed. The twine material also has an effect as has the accumulated weight and volume of fish in the cod end at any one time.

A further point about mesh size regulations concerns the ability of resourceful fishermen to get around them. Let us take a hypothetical case of a bottom trawler working for demersal fish in the North Sea. In his area the minimum mesh size is say 80 mm. This is fine for haddock and cod, but suppose the vessel encounters quantities of whiting, most of which can escape through an 80 mm mesh. What can the skipper do if he wants to get around the regulation? He could (a) change the cod end temporarily; (b) add a blinder or second bag and cod end over the first; (c) restrict the meshes in some other way so they do not open fully. Unless he was apprehended in the act of fishing with either modification, it would be extremely difficult after the event to prove any infringement of regulations. There are in fact ways of accomplishing (c) which would probably not represent a breach of regulations at all, e.g. doubling the number of meshes in the cod-end circumference.

CONTROL OF THE AMOUNT OF FISHING BY CATCH LIMITS

The assumption on which the present Consultation is based is that a country (or a group of countries sharing a stock) has determined that in relation to national policies (economic efficiency, greater food production, etc.) the amount of fishing on some given stock is too high and should be reduced.

The simplified rationale behind such a decision, which is a typical yield curve as given in numerous volumes and reports on fishery management. This gives the relation between sustained catch and number of standard vessels. The relation between the number of vessels and the abundance (biomass) of the stock. Without any regulation the economic long-term equilibrium is at the point A, where 125 vessels are catching 56,000 tons. A better state (according to the policies of the country) would be B, where, in the long-term, 76 vessels would catch 60,000 tons. Apart from an increased total catch, these 76 vessels should generate much profit, which could go into the pockets of (a limited number of) deserving fishermen, be siphoned off to central government or used in other suitable ways.

In principle there are several ways in which it can be determined that the fishery is being maintained close to the target point (e.g. by checking that the stock abundance is around 90,000 tons, or that only about 70 standard vessels are operating), and several measures by which the necessary controls can be applied (e.g. catch quotas, licence limitations, or various indirect measures). The same measure need not be used at all stages in the process. For example, it may be decided that the best long-term target is to maintain the spawning stock at 90,000 tons; that to do this the catch in 1983 should not exceed 60,000 tons, and this will be achieved by imposing a tax of $50 per ton on all landings, thus discouraging all but a few vessels from going fishing (the level of the tax being set so that just that number of vessels will consider it worth going fishing as is required, given the likely catch rates in 1983, to catch 60,000 tons).

If life were as simple, then the choice of measure for setting targets or controls would not be critical. Unfortunately, for virtually no fishery an adequate representation of biological reality, nor are the commercial and administrative facts of life such that deciding that the quota for 1983 should be 60,000 tons is sufficient to ensure that no more than 60,000 tons are caught. Among the few stocks is a reasonable (though not exact) representation of biological reality are whales. The history of whale management does not suggest that biological simplicity alone is a guarantee of successful management.

The choice of the appropriate measures (catches, effort, etc.) for setting targets and controls can be critical. Further, because the nature of the biological, economic, social and administrative complications vary greatly from fishery to fishery, the best measure is not the same in all fisheries but for each fishery has to be carefully chosen after looking at the particular condition of that fishery.

Nevertheless, a superficial examination of fishery management today would suggest that the only proper way of managing fisheries is through catch quotas. The apparently interminable discussions within EEC are concerned very largely with the value of Total Allowable Catches (TACs) and how they should be allocated. Some of the key phrases in the draft convention of the Law of the Sea - "The coastal state shall determine the allowable catch" (Article 61(1)) or "Where the coastal state does not have the capacity to harvest the entire allowable catch, it shall ... give other States access to the surplus ..." (Article 62(2)) - imply that catch limits are the key to good management. This is far from being the case; later sections in this paper will show that in many situations attempting to control the fishery through catch limits is a second - or third - best method.

Also, as shown in another paper, the so-called "surplus" is like a unicorn - very nice in theory and in illustrations of the ideal life but seldom seen in real life. (These comments are, of course, no criticism of the UN Conference on the Law of the Sea, in which fishing was a relatively minor item. Within national delegations the fishery experts were concerned principally in ensuring that their national interests were not hurt. For this purpose meaningless or indeterminate phrases often served well. The only ones among those attending UNCLOS whose prime interest was to see that the text provides as good a framework and guide as possible for rational fishery management and development were some of those from international organizations. These were attending as observers and were relatively powerless to influence the outcome.) To understand why catch quotas have achieved their current prominence, and thereby make it easier to come to a fresh evaluation of the relative strengths and weaknesses of different approaches, it is necessary to give a short historical review.

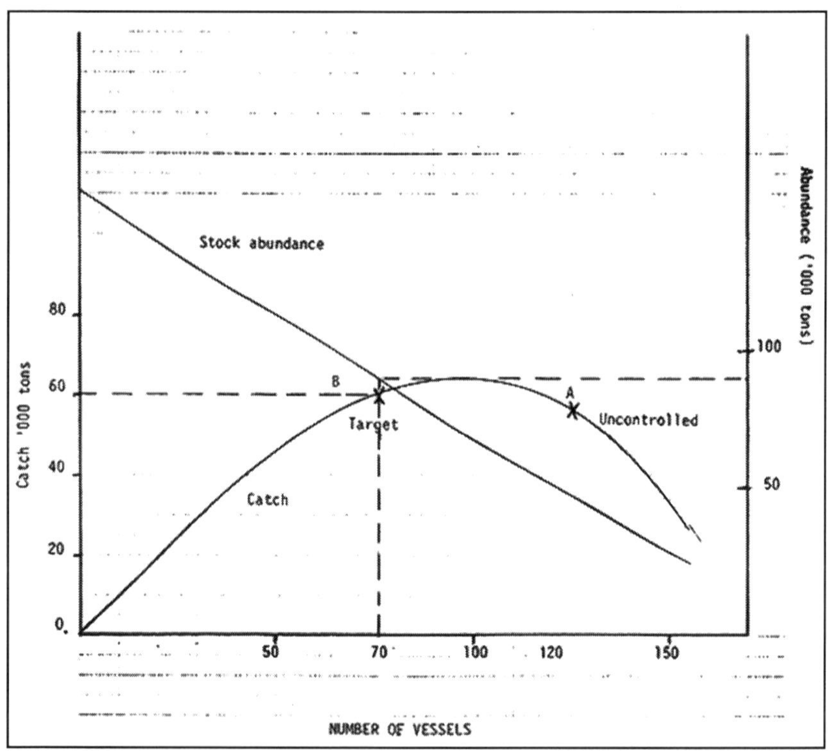

SOME HISTORY

Regulation of important commercial fisheries by means of catch quotas goes back nearly half a century to the North Pacific halibut fishery (Skud, 1973) and, immediately after the second World War, Antarctic whaling (Allen, 1981). However, the key period, when the attention of fishery scientists and administrators became focussed on the control of fishing effort, and when it became commonly accepted that the proper way of doing this was by catch quotas, was in the 1960s, and especially in the two North Atlantic Commissions (NEAFC and ICNAF).

By the 1960s, regulation of the size of trawl meshes had been in force in several parts of the North Atlantic for some time. It was clear that while such regulations did give benefits, there were limits to how far they were effective in multi-gear, multi-species fisheries, i.e. the conditions in most of the area except some of the most northern parts (Bear Island, Labrador) which were nearly pure trawl fisheries for cod. It was also clear that such regulations could do little (and then only in the short- to medium-term) to tackle the economic problems of excess effort and hence excess coasts. ICNAF in particular was active in promoting basic studies of the problem (e.g. Templeman and Gulland, 1965; ICNAF, 1968). These studies showed that there were two basic choices that need to be made: (a) whether the amount of fishing should be measured and controlled in terms of catch or effort (e.g. number of days at sea), and (b) whether there should be a single global quota or whether there should be allocations to countries of specific shares in the total quota.

The second question was easily answered. Both theoretical studies and practical experience in the fisheries for whales and halibut showed that without allocation there would be a scramble for the quota (e.g. the "Whaling Olympics") and most, if not all, the potential economic benefits would be dissipated. Though a single unallocated quota still continues to be used in some fisheries, e.g. the Eastern Pacific yellowfin tuna, horse mackerel in the South-East Atlantic, these are exceptions. In practice the system of exemptions used in the tuna fishery meant that there were de factoallocations, with the amount left after the exemptions were dealt with being in effect the US allocation. There are doubts about the effectiveness of the ICSEAF measures.

To answer the first question - measuring the amount of fishing in terms of catch or effort - it was therefore necessary to look at the implications for allocations, as well as the other factors such as the constancy of the relation between catch or effort on one hand, and fishing mortality on the other. In the conditions of the North Atlantic Commissions this tipped the balance very clearly towards using catch.

Catch allocations are easily understood; if one country has a quota of 50,000 tons and another 100,000 tons, the second is, in at least one obvious sense, getting twice the benefits of the first. Whether the real benefits are twice as much is another matter, depending on the ways in which the countries use their quotas and manage their fisheries to reduce costs and increase the gross returns (e.g. by timing the landings so as to get the best prices). Everyone concerned - the ordinary fisherman as well as the administrators round the negotiating table - could see what was happening to their shares, in absolute terms and relative to others. In contrast, with the variety of vessels and gears involved, effort allocation was viewed as complicated and difficult to understand.

It is far from clear what combination of numbers of Portuguese dory vessels, English wet-fish trawlers, and Russian factory ships would be needed to achieve a target fishing mortality and a fair allocation between countries.

Further, and this was a point given some emphasis in the North Atlantic Commissions, the efficiency of each of these groups will change with technical improvements, and change by different amounts. Thus the target efforts, in total and for each country, would have to be continuously changed. It would be difficult scientifically to determine the necessary change but even more difficult, politically and administratively, to get the necessary changes agreed to, especially to get a country whose fleets have increased their efficiency a lot to agree to a greater than average reduction in its allocation in terms of nominal effort.

It was recognized that the target catch would have to be altered from year to year to allow for changes in stock abundance (e.g. presence of a good year-class). However, it was felt that these adjustments - that did not have to involve differential adjustment to national allocations - could be dealt with. In the context of the main fisheries being discussed - principally the northern cod fisheries - this was reasonable. In most of the stocks the fish are long-lived; typically fish recruit at three to five years old, but do not make their biggest contribution to the fisheries until perhaps six to eight years old, and some of the fisheries on spawning fish depend on even older fish. Thus the strength of most year-classes in the fishery could be estimated from their occurrence in the commercial catches of previous years. The exceptions - the one or two youngest year-classes - did not contribute significantly to the catches and even these, it was believed, could be estimated if necessary by research-vessel surveys of the pre-recruit fish.

To a large extent these expectations have been fulfilled. Allocation of national shares of some Total Allowable Catch has been established as a workable method of controlling the total amount of fishing in a multi-national fishery. Scientists have found it possible to agree among themselves on the values of the TAC to be set each year and these recommended values of the TAC have been, at least in a good number of cases, approximately correct in terms of the state of the stocks and the objectives of the managers. The fishing industry and administrators have learnt that it is convenient to follow the recommendations of the scientists on the value of the TACs.

There have been exceptions. The scientific estimates of the appropriate TAC have sometimes been in error, and the countries have sometimes been able to reach agreement only by agreeing on a TAC in excess of the scientific recommendations. On occasion these exceptions have been serious, e.g. in relation to some herring stocks (Saetersdal, 1980). One cause of scientific error is relevant to the present discussion because it arises from control by catch, rather than effort. To calculate TACs it is common practice to use Virtual

Population Analysis (VPA). In this the choice of the terminal F (i.e. the fishing mortality assumed to be occurring in the last year for which data are available) is critical (Pope, 1972). If, in say 1982, when the TAC for 1983 is being calculated on the basis of 1981 data, the value of F for 1981 is under-estimated, this means that the stock in 1981 will be less than one thought and the catches during 1981 will take a higher proportion than expected of this smaller stock. Thus the degree of over-estimation of the stock at the beginning of 1983 (and of the TAC for 1983) may be considerably greater than the original degree of under-estimation of 'terminal F'.

However, even with these exceptions, the events of the past years have confirmed that catch quotas were appropriate to the conditions of the multi-national fisheries. Nevertheless, the extent to which attention of managers and others have been focussed on catch quotas, and specifically on determining the value of the TAC and its allocation, has had two serious effects.

The first effect has been that within the North Atlantic fisheries (or at least in the north-east), the TACs have become ends in themselves. The general impression given by much of what is said about current problems of managing North Sea fisheries is that if only the correct TACs were agreed on, and complied with, then the fisheries would be properly managed. (This also appears, with a nationalistic flavour, in the belief that if all the foreigners would accept a proper, low share of a TAC, then all would be well.) In fact a catch quota is only a means to an end; unless further action is taken, the imposition of a catch quota goes only a little way towards reducing excess costs and realizing the large economic benefits pointed out by the ICNAF studies of nearly two decades ago. This is not the place to attempt an analysis of what is wrong with current management practices in the North-east Atlantic, still less to suggest prescriptions for improvement. However, it does seem that some shift in attention away from catch quotas could be useful.

The second effect has been more widespread. This is the belief that catch quotas are the only correct way to manage a fishery. Nothing in North Atlantic experience, or in the arguments put forward in favour of catch quotas for the North Atlantic cod fisheries of the 1960s, prove that they would be the most suitable way of controlling the amount of fishing in most present-day fisheries. One factor favouring catch quotas (that a large number of countries all had free access to the stocks) now only applies to tuna and a few other species that can be taken on the high seas. Another (that the stocks were of long-lived fish, without large and unpredictable changes in abundance from year to year) is not typical of most other resources, especially in warmer waters.

COMPLICATING FACTORS

The assumption an adequate representation of real life and hence that, from the resource point of view, the choice of management measure is unimportant, can break down in various ways. This can lie within the model itself or can lie outside the model.

Within the model it is not necessarily true that fishing mortality, fishing effort and catch have a neat one-to-one relation. In different years the same catch or the same nominal fishing effort will give rise to different values of fishing mortality. If, as is common, the basic strategy of management involves maintaining a certain fishing mortality, this is undesirable and can be dangerous. Similarly, different catches or different levels of fishing effort (or fishing mortality) will be needed to maintain a certain spawning stock, depending on the current abundance. An annual catch quota must therefore be based on a reliable and up-to-date estimate of the abundance of the total stock, and as necessary of the spawning stock.

The assumptions of the economic aspects of the model, that the costs and value of the fishery are proportional to the fishing effort and the weight caught, respectively, are not always true. If they are not, then any changes in the pattern of fishing brought about by management will not have the economic effects predicted. To the extent that the departures from the assumptions are actually caused by the management action, the results can be very significant. This is particularly true when, as is usually the case, the effect of the regulations is either to increase the costs of a given level of fishing effort (or, more correctly, of exerting a given level of fishing mortality) or to reduce the value of a given catch. When managing an over-exploited fishery, with excess catching capacity, it is often easiest to achieve the immediate objective of reducing the fishing mortality by reducing the efficiency of the fleet (shorter open seasons, prohibition of certain types of gear, etc.). This will increase the costs of exerting the target mortality. Any disruption of the normal pattern of fishing will also tend to reduce the value of a tonne of fish, e.g. by concentrating all the landings in a short open season.

The simple model ignores many biological complications even within a unit stock. All fish in the stock are not the same. A catch of a ton of fish will have different effects on the stock and fetch a different value on the market depending on the size or condition of the fish. Very small fish may be worth so little that they are discarded at sea. Management measures may cause the sizes or conditions of fish caught to vary and may change the discarding practices of the fishermen.

No fish stocks exist in isolation, and many fisheries are concerned with only one stock. Biological interactions, e.g. between predators and prey (whales/krill, cod/herring, seals/capelin), have seldom been explicitly taken into account in setting management measures in the past. In the future, though, it is clear that they will have to be more carefully considered. This raises difficult biological problems, e.g. in designing suitable models and still more in fitting the models to data. They also raise difficult social and political problems, especially when marine mammals are concerned, in determining the most suitable management policy, e.g. in balancing the interests of those wishing to harvest krill against those wanting the most rapid rebuilding of whale stocks. These problems concern the strategy of management, e.g. determining the appropriate levels of fishing mortality for the various interacting species. They however do not affect the question at issue here, of the correct tactics and the measures for achieving, e.g. the target fishing mortality.

The tactical questions are the extent to which one or other choice of measure to manage one stock can affect the harvesting of other stocks. Preferably, the controls on an over-exploited stock should encourage the greater use of under-exploited stocks. The practical effect is however often the opposite. Restrictions on the type of the gear that can be used are particularly likely to inhibit the greater exploitation of under-used stocks. Difficult problems are also posed by the incidental catches of the "protected" species by vessels fishing for other species - should they be allowed to keep them and if so how many. A low limit can mean loss by discarding and a high limit a degree of fishing that is really directed at the target species.

As more management schemes are implemented, the incidental costs of different patterns of management and some of the unexpected effects, not covered by simple theoretical models, are becoming apparent. The most obvious costs are of the additional research, administrative and enforcement activities that are needed to determine the regulations for each year in the necessary detail and ensure they are actually followed. Enforcement, particularly when it requires close checking on the activities at sea, can be very expensive. These costs, and also the costs of research, can be greatly affected by the type of measure used.

The most serious indirect effect of regulation occurs when they lead to mis-reporting of basic data. Where catch quotas are in operation but not fully enforced, the actual landings may be considerably in excess of the figure reported. This has, in some cases, reached the stage where the scientists concerned have had to report that it is no longer possible for them to make meaningful assessments of the state of the stock.

The degree to which management by catch quotas can deal with these complications, as judged by the four criteria of maintaining the resource, economic efficiency, equity, and transaction costs. First, though, the actual mechanics of applying a catch limit will be examined.

THE MECHANICS OF CATCH LIMITS

The implementation of a limit on catches involves two or three inter-dependent stages - determination of the level of the total allowable catch (TAC), usually though not always the allocation of shares in the TAC to different groups of fishermen, and some control system to ensure that the catches of each group are no more than their allotted share. Normally, the common practice is for the TAC to be set according to the scientific recommendations. These will be based on some interpretation of the basic management objective (to exert a fishing mortality equal to $F_{0.1}$, to maintain the stock at the level giving the MSY, etc.), together with the available information on the current stock abundance, etc. In practice the TAC is often modified from the scientific conclusions in order to resolve some immediate problem.

The preparation of the scientific recommendations, which for the ICES area has been described by Hoydal in a paper to this Consultation, meets two major obstacles. First, the value of the appropriate quota is not uniquely determined by the biological conditions but depends on the specific objectives (MSY, $F_{0.1}$, etc.). These may not be explicitly defined and the scientists will then have to make some assumptions. Second, the data on the current conditions of the stock will never be absolutely precise. Even if pre-recruit surveys are carried out, the strength of the youngest year-class in the stock in the coming season will rarely be accurately known. There will therefore always be some doubts and imprecisions in the estimated quota. Only rarely is there a good understanding between the scientists and the managers on how to deal with these uncertainties.

If the scientists take the natural course and express their conclusions as a range - "to exert a fishing mortality equal to $F_{0.1}$, the quota in 1983 should be between 95,000 and 120,000 tons", then too often the administrators have taken the easy course, and used the upper limit (i.e. 120,000 tons). The response of the scientists has then often been to play down the uncertainties, and to give a single figure, usually towards the more conservative end of the range. In doing this they are tending to act as though the scientists had the ultimate responsibility for the resource. In the short run this is probably a good thing, with better levels of TAC being set. In the long run, however, it would seem more appropriate for the uncertainties in the scientific advice (which include doubts about the models being used, as well as the estimates of year-class strength, etc.) and in the models to be dealt with in the final decision-making process together with all the other uncertainties.

The allocation of the TAC between various groups is the key stage so far as economic efficiency and equity is concerned. Without some allocation, implementation of a catch quota involves an ever-shortening open season, during which there is an open competition between all the fishermen to get as much as they can while the season lasts. The chances of much economic benefit is small. In international fisheries the importance of allocation to countries is well recognized.

Allocation is equally important within national fisheries, until the sub-division of the fishery is fine enough for there to be little competition within a group. How much division is needed will vary. Antarctic whaling was carried out by large expeditions, so that even at its peak each national quota had to be shared by at most two or three big companies. Near-shore fisheries, dominated by small family-owned boats, may require several layers of allocation - perhaps first by port, type of gear, or size of vessel, etc. and then sub-divisions within these categories.

Allocation within a country need not be expressed solely in terms of catch, even when the overall limit is so expressed. Limits, for example, on the total number of vessels that can be operated, or on the numbers operated by a particular company, will prevent much of the uncontrolled competition that dissipates the benefits of an un-allocated total catch quota. Such forms of allocation can avoid some of the problems that arise when all allocations are made in terms of catch. Then there can be a lack of distinction between allocation of a proportional share (in some unspecified total) and a share expressed as so many tons of fish. The latter runs into trouble if the stock should decline, e.g. due to poor recruitment; then the sum of the allocated shares may exceed what can be taken from the stock.

The problems of enforcement of catch quotas have, until recently, been neglected, at least in the published literature. When there was little detailed allocation, it might not unreasonably be assumed that catches were correctly reported. The problems were those of having a sufficiently quick administrative system, so that fishing could be stopped rapidly once the target catch was approached. Once each small group is allocated a share in the total, then there will be an incentive for the fishermen in that group to under-report their catches. This mis-reporting, which if extensive, can undermine the whole statistical system on which the assessment and management of the fishery is based, has apparently reached the stage in some areas where the scientists have to state that they can no longer make reliable assessments of the state of the stock.

EVALUATION

Effect on Resources

In principle, given adequate knowledge of the abundance of the stock in the coming season, a catch quota enables the impact of the fishery to be maintained at the desired level, whatever that may be. In practice, knowledge of the resource is less than perfect and the catch quota set will not be the correct value to achieve the desired input on the stock. The extent to which its effect will be closer to the desired one than those obtained through other types of control (e.g. on effort) depends on the year-to-year variability of the stock, and the degree to which these can be measured and predicted. At one extreme, whales live a long time and the population varies little. Catch quotas are highly suitable for whales. Stocks of long-lived fish (halibut, cod in northern waters, etc.) live a long time and, though year-classes vary, it is possible to monitor these changes reasonably well in advance of their full impact on the commercial catches. For these too, control by catch is suitable, though probably no better than effort limits. For the more short-lived stocks, e.g. those in the North Sea, and especially in tropical areas, it is difficult to make accurate estimates of the abundance in the following season. A significant part of the catch comes from the newly recruited year-class which can only be estimated by pre-recruit surveys.

These tend to have both high costs and high variance. For these stocks other types of control, e.g. on effort, may provide better control of the resource. Exceptions are some of the shoaling pelagic species. Because for these the catchability coefficient varies inversely with abundance, limits on the nominal fishing effort gives little control of the actual fishing mortality. On the other hand, it may be possible to use acoustic surveys at the beginning of the season to get a direct estimate of the current abundance, and thus use catch limits successfully.

Catch limits are useful only if enforced. Enforcement is however almost inevitably of landings, not catches. Some forms of "catch" control can lead to increased discards. The landing of "regulated" species (e.g. Pacific halibut) by fisheries directed to other species may be prohibited. Alternatively a certain percentage of by-catch may be permitted. This can lead to deliberate fishing for the "regulated" species early in a trip to make sure of the allowed percentage, followed by the discarding of any further catch. To the extent that such discards are poorly recorded and not taken account of in the statistical data, the impact on the resource is greater than expected.

Economic Efficiency

This depends on the extent to which the total catch is allocated to individual groups. At one extreme, with a single global quota, management leads to a scramble for the quota, a short season, and great economic inefficiency (excess capital employed, high catching costs, and often reduced value of the catch). At the other extreme, when each individual enterprise is allocated its own share, economic efficiency is high. If a fisherman knows that he can take say, 100 tons at some time during the year, he can arrange to take this at the most profitable time taking account of seasonal variations in fish conditions and market demand, the costs of catching, and the opportunities to engage in other fisheries.

Compared with control of effort, catch limits - provided that there is some allocation - have the economic advantage that they do not lead to distortion of fishing practices as fishermen try to out-manoeuvre the manager, e.g. by building larger boats, or more powerful boats within a tonnage limit. These distortions must inevitably lead to some degree of inefficiency.

Equity

This too depends on the nature of the allocation rather than catch limitation per se. Without allocation, not only will much of the economic benefits be dissipated, but such benefits as are gained will go to those who can maximize their share of the catch. These will be those fortunately located with respect to fishing at the beginning of the season, or who can best build up their catching capacity. These will seldom be those who the managers would actually wish to give particular preference to.

When there is allocation, the pattern of allocation can obviously be arranged so as to favour any desired group. With detailed allocation the social impact of control by catch is likely to be similar to that of control by effort. Similar problems will be raised - who should get licences/quotas, for how much, can they be transferred. An apparent advantage of catch limits is that they are clearly comparable - a man with a quota for 100 tons has twice the "benefit" of someone with 50 tons. This is certainly an advantage in multi-national fisheries. It may be more apparent than real at the national level, where the question is whether or not an individual can continue to run his boat, or whether a company has to reduce its fleet, and by how much.

Transaction Costs

The year-to-year adjustments in catch limits, which are essential in all but the least variable of fisheries, are costly. At the least, they involve research staff in time-consuming re-calculations, often under severe pressure of time (dealing with this year's data in order to set next year's catches). The costs in terms of diversion of scarce scientific man-power from other tasks, including the adequate consideration of the broader aspects of resource assessment and management advice, may be large. Often there will also be considerable financial costs to carry out, for example, pre-recruit surveys. These scientific costs are likely to be more than those involved in other forms of control.

In contrast, to the extent that controls on the amount landed (but not the amount caught) can be checked in part, costs of enforcing catch regulations are at least no higher than those of enforcing other measures. However, if several stocks of the same species are being managed independently, and the fish from different stocks cannot be distinguished on the market, then control must be carried out at sea.

Also, particularly for high-priced species (lobster, salmon, sole, etc.) it is possible for much of the catch to be sold without passing through recognized markets (or to pass through those markets identified as something else). Control of catch limits may therefore not be as easy as might be supposed from a superficial examination, and may well involve costly control at sea. In addition, it may be less obvious to other fishermen that catch limits are being observed than limits of other kinds - it will be fairly clear what vessels (and of what size and type) are fishing in an area, but less clear how much they are catching. It may be harder in the case of catch limits than for effort limits to build up a mutual trust between fishermen that the regulations are being observed by all those in the fishery, although this trust may be the key to reliable implementation of management measures. Finally, it should again be noted that poorly enforced catch quotas can undermine the supply of reliable catch statistics, which is the basis of the scientific side of fishery management.

REGULATION OF FISHING EFFORT THROUGH VESSEL LICENCES

The harvesting of all important fish stocks in Norwegian waters is restricted by quotas, i.e. total allowable catch (TAC). For some species, TACs are determined solely by Norwegian authorities; for others this is done in cooperation with neighbouring countries. In cases where the fishery is stopped when the quota is exhausted, which is the case in most Norwegian fisheries, vessel licences become invalid because of fear or over-exploitation. What purpose does licensing have then? The question is appropriate as licensing systems (limited entry) still play an important role in Norwegian fishery policy.

The original argument for imposing licensing schemes has differed from case to case. I shall give you two examples:

Example 1: Restricting trawling for cod through a licence scheme (1936) had at least three purposes: (i) protection of the coastal fisheries; (ii) large and costly trawlers meant that foreign capital gained entry into the fishery; (iii) trawling, it was said, was a threat to juvenile cod.

Today these arguments have only limited validity. The catches taken by Norwegian trawlers do not influence to a significant degree prices on the world market for fish. The extended mesh size in the cod-trawl (minimum 135 mm from 1 January 1983 in the Norwegian zone north or 64°N) means that trawling no longer has the same negative effect on the young cod. In addition, since 1974, cod-trawling has been regulated through a separate quota fixed for trawlers. Even without licensing, the catch taken by trawlers would remain at the same low level.

Nevertheless, a licence scheme is maintained. Why? (1) Since there is excessive capacity in relation to the quotas fixed for trawlers, the situation would be even worse if new trawlers were allowed to enter the fishery. Present economic conditions, however, make the possibility of new projects seem remote anyway. But conditions often change and licensing is maintained as a form of “insurance”. (2) The geographical distribution of licences provides an instrument to ensure a certain supply and distribution of raw fish to regions that are particularly dependent on fish as raw material for processing.

Example 2: The licence scheme for the purse-seiner fleet was implemented in 1973 following a remarkable increase in catching capacity from 1970 to 1972.

The central authorities feared over-exploitation of the most important stocks. It was also felt that, if the high rate of investment in this sector were to continue, economic disaster would follow, either because of depleted stocks or (if quotas were to be implemented and respected) because of severe over-capacity.

Biological considerations are no longer an important argument for imposing the licence scheme, however, because every year since 1977 a total quota (TAC) for capelin, which is the most important stock for the purse seiners, has been negotiated with the Soviet Union. As anticipated, quotas have meant that over-capacity would be a prominent feature of the purse-seiner fleet. Nowadays the licence scheme for purse-seiners has two main purposes: (1) To improve the economy of this fishery by limiting the number of vessels involved in taking the quotas. (In fact, the number of vessels is steadily decreasing, since concentration of the cargo capacity of two small or medium size vessels in one larger one is permitted.) (2) The cargo capacity registered by the licence is the basis for dividing the capelin quota between the participating vessels.

Many fishery politicians refer to a third purpose for the licensing of purse-seiners: that is, to influence the regional distribution of such vessels, in order to allocate job opportunities to coastal areas which seem to have greatest need for them. The possibilities for steering the geographical distribution of licences, however, are in practice very limited. One argument for having a licensing arrangement is of a more political nature: it is far easier to enlist fishermen's support for the necessary quota regulations when the fishing fleet is not expanded excessively in relation to the proposed quotas.

On account of protests from fishermen, substantial over-capacity in relation to the national quotas may force the fishery authorities to seek increased TACs (beyond the level recommended by biologists) when international negotiations over TACs take place. If over-capacity prevails in several countries, the pressure for higher TACs may become dangerous. This can be avoided to a certain degree by implementing limited entry. For the period 1979-82, the Norwegian Government adopted a scrapping programme for certain groups of fishing vessels, in an attempt to get rid of over-capacity. The purpose was not only to make it easier for the politicians to fix more restrictive quotas, but also to lower costs in fishing. This would reduce the need for government support. The scrapping programme included only vessel groups subject to limited entry. Without entry control, new vessels could be expected to replace the capacity withdrawn through scrapping. Licensing thus has a legitimate use also in cases like this.

Still important groups of vessels in the Norwegian fisheries are not subject to limited entry. The fleet of small and medium-size coastal vessels operating in the cod fisheries in northern Norway is regulated through other measures such as a ban on fishing during certain periods spread over the year. This fleet is also subject to gear control (maximum number of nets per vessel). The purpose of these measures is to reduce the catch to the level fixed on the basis of international agreements.

This illustrates that we are able to restrict fishing effort by administrative means other than licensing; but, if our objective is to restrict the amount of production factors involved in taking the quotas, in order to increase profitability, licensing schemes can be an appropriate instrument.

Can Limited Entry Replace a Quota System?

I shall now turn to the following question: can a comprehensive licensing system replace a quota system? Is it possible for a licensing scheme to achieve biological (conservational) objectives in addition to the economic and distributive objectives already mentioned? The answer is that, for stocks which might migrate between the zones of several nations, licensing, even when used by all relevant nations, is not sufficient to avoid depletion of such stocks. The nations must in addition cooperate and agree on a TAC and on the sharing of it.

If we know our national share of the TAC, however, can licensing then be an appropriate instrument to avoid over-fishing of the national quota? The same question of course applies when we consider stocks found within the zone of one nation only and solely fished by vessels from that State. My answer to this question is: "Yes, it is possible to use licensing as a measure to keep the catch within national quotas; but, at the same time, it is far from an advisable policy if not combined with measures of an economic nature".

I shall give you some arguments in support of this view:

1. The typical case is that a licence scheme is imposed after a stock has been too heavily harvested, in order to prevent its further depletion. The introduction of a licence scheme normally means that licences are issued to the vessels already active in this fishery but not to newcomers. Restoring the stock, however, requires reduced fishing effort. To achieve this (disregarding the auctioning of an appropriate number of licences), you must introduce a system of quotas, causing some or all of the vessels to be laid up for certain periods of the year or transferred to other (less profitable) fisheries.
2. Even if the licence scheme is implemented at such an early stage that over-utilization of the stocks has not yet occurred, continuous technological progress will increase the catching capacity of the fleet. Sooner or later even the restricted number of vessels will be able to deplete the stock or over-fish the national quota.
3. TACs and national quotas will often change from year to year, depending on the stock situation. The catching capacity of licensed vessels, however, obviously will not change in conformity with changes in the recommended catch volume. Instruments other than

a licence system, therefore, are needed. The same kind or argument is valid even if the national quota remains stable, as the catch rate may vary from year to year irrespective of the size of the stock.

4. A licence scheme has several disadvantages. It retards the flexibility of the fleet, for example, reducing its ability to adapt to new conditions. It also renders the recruiting of new and young fishermen into the fishery more difficult.
5. The management of a licence scheme is often a vast administrative under-taking. One well-known difficulty is the handling of applications for replacing a licensed vessel with another, better suited (and always a little bit larger) vessel. I may also mention the questions which arise in connection with buying and selling licensed vessels. Should a licence be transferable?

So much for the disadvantages and shortcomings of a limited entry system. Combined with quotas and monetary measures, however, a licensing system may be very appropriate. A levy on catch or a licence fee, in some cases combined with a buy-back programme, may possibly solve most of the problems mentioned.

Licensing: Still a Necessary Supplement?

Licensing systems are still considered to be very important instruments in Norwegian fishery policy. The Fishermen's Association has even suggested that all types of fishery should be subject to licensing as a means of regulating the fisheries. In recent years, however, the central authorities have become more aware of the many disadvantages connected with licensing as a regulatory measure. In addition, the original purposes for imposing the several licence schemes are nowadays to a large degree served by an array of new instruments, mainly quotas.

As long as over-capacity prevails, however, and the level of investment is not controlled in one way or another, the licensing programmes are instruments which may be convenient to have at hand in order to restrict increases in capital and running costs. In addition, I may remind you of the convenience of implementing a licence scheme before the introduction of a scrapping programme, and of the pressure for increased TACs that may arise when catching capacity is not restricted through licensing.

Licensing means that administrative bodies make decisions which would otherwise have been determined by market mechanisms. Were it feasible to consider a licence as a commodity, subject to buying and selling, the serious disadvantages connected with licensing probably could be avoided.

7

Characterization of Fishing Gear Technology

Most studies on the effects of fishing gear on the seafloor have been focused on trawl and dredge fisheries. A significant portion of the landings of finfish and shellfish from U.S. coastal waters bottom is made with contact, mobile fishing gear. Bottom trawls catch mostly groundfish and shrimp; dredges are used primarily for scallops and clams. Because different gear types cause different types of seafloor disturbance, it is important to understand the nature of the gear used in the fisheries before considering the effects of those fisheries on habitat. Other gear, such as pelagic trawls and seines, are not included in this chapter because they either have less contact with the seafloor or they represent a minor fraction of the total fishing effort.

Fishing gear that is dragged over the seabed or through the water is called mobile gear. All bottom-contact, mobile fishing gear disturbs the seafloor to some extent. The dragged gear generally includes a bag constructed of synthetic webbing or metal rings and chain links that collect the catch. It is classified as active fishing gear because the animals do not voluntarily enter the gear; they are either swept up from the seabed or netted from the water by the gear. As fishermen strove to fish more efficiently, dragged gear evolved to cover greater areas, or sweeps, of the seafloor. Dragged gear initially was deployed from handpowered boats, then sailing vessels, and finally from diesel-powered ships, some with engines greater than 1000 horsepower (HP).

Mechanization of fisheries with powerful engines and winches allowed larger gear to be towed faster and handled with less labor. Improved vessel design made it possible for fishermen to travel farther from their home ports and to fish under adverse weather conditions. Improvements in fishing vessel electronic navigation equipment from radio direction finders (LORAN), collision avoidance equipment, radar, and finally satellite-based navigation, such as global positioning systems (GPS), have increased fishing efficiency. The application of sonar technology to fish finding and to mapping the seabed has improved the ability of fishermen to target areas of the seabed that were not previously accessible to dragged gear. Technology also has helped fishermen avoid sites where rough seabed terrain could result in the loss or damage to gear. Dragged gear has developed from small shellfish rakes towed by hand-paddled canoes to groundfish trawls that are equipped with net mouth openings that in some cases exceed 50 × 100 m.

Trawling off the United States began in the early 1900s, and at about the same time on the Atlantic and Pacific coasts (between 1901 and 1905). By 1971, trawling was considered "a most efficient method of offshore fishing" (Shapiro, 1971). Trawling extended rapidly along the Atlantic coast and into the Gulf of Mexico and became the dominant method of catching fish and shrimp for human consumption. Although trawl fishing also spread along the Pacific coast, the relatively narrow continental shelf and limited West Coast markets led to establishment of a smallboat fishery (generally less than 25 m in length) until the advent in the late 1960s of foreign fisheries.

After World War II, distant water foreign trawl fisheries developed off the New England states, along the Pacific coast, and in the Bering Sea and Gulf of Alaska. Those foreign fisheries were largely phased out by 1990 with the implementation of the Magnuson-Stevens Fisheries Conservation and Management Act (1977), which extended U.S. jurisdiction over fishery resources throughout the newly declared 200 mile exclusive economic zone. Large-scale domestic catcher–trawler and mothership operations arose in the waters off Alaska (Ianelli and Wespestad, 1998) and from the coast of northern California to Washington. In New England, the trawl fisheries also expanded as larger, more efficient vessels were introduced into the fleets. Currently, there are major groundfish trawl fisheries in New England and Alaska. The shrimp trawl fishery dominates in the Gulf of Mexico and in shallow waters off the coasts of the Carolinas and Georgia. Groundfish and shrimp trawl fisheries operate off the coasts of Washington, Oregon, and California.

DREDGE GEAR

Most dredges are rakelike devices that use bags to collect the catch. They typically remove molluscan shellfish from the seabed, but occasionally are used for crustacea, finfish, and echinoderms. Dredges take either epifauna or infauna; the design details of the gear are fauna specific. On soft bottoms, the dredge flattens the microrelief on the seabed (wave-ripples) and resuspends fine sediments. On hard rocky bottoms, the dredge scrapes off epibenthic organisms and disturbs the substrate.

In estuarine waters, dredges are used to collect clams, oysters, conch, and crabs. The oyster dredge consists of a steel frame, 0.5–2.0 m wide, with an eye and "nose" or "tongue" and a blade with teeth. The tow chain or wire and a bag for the catch are attached to the frame. The dredge is towed slowly (<1 m/s) in circles, from vessels that are 7–15 m long. Similar dredges are used to catch blue crabs in the mid-Atlantic region during the winter. Stern-rig dredge boats (averaging 15 m long) drag two dredges in tandem from a single chain warp.

The dredges are equipped with long teeth (10 cm) that dig the crabs out of the bottom. The same gear is used in the Chesapeake Bay to catch whelk in summer and mussels in fall.

In the soft clam fishery, which occurs in shallow estuarine waters, the dredge head (manifold and blade) is attached to a conveyor or belt that carries the materials retained on the blade to the working deck of the vessel. These dredge vessels (generally about 15 m long) are restricted to shallow water—less than one-half the length of the escalator. Because soft clams live in shallow waters (2–6 m), this poses few limitations for the small vessel operators who use this gear.

Offshore, large dredges catch sea scallops, which inhabit a sand–gravel–cobble bottom and live on the surface of the seabed. Because scallops sense and retreat from slow-moving dredges, scallop dredges are towed at speeds of up to 2.5 m/s. The scallop dredge has a steel frame with a tongue with an eye, a blade with no teeth, and a bag. The width or mouth opening of the dredge ranges from 3.0 to 4.5 m, and dredge weight varies from 500 kg to 1000 kg. The largest scallop dredge vessels, about 60 m long drag two 4.5 m dredges, one from each side of the vessel, and they use winches and navigational electronics to maintain high efficiency. Scallop dredges disturb the seabed. Disturbance of the seabed by dredges is required to dislodge scallops for capture in the net.

Special methods are used to gently extract clams that burrow in sediments. Water jets are directed into the seabed to liquefy the sediment suspending the clams up in a sediment slurry. This blade sieves the clam– sediment slurry. The hydraulic clam dredge is used offshore to collect surf clams and ocean quahogs and inshore to collect soft clams. Hydraulic dredges are efficient, but they restructure sediments and disturb sediment biota.

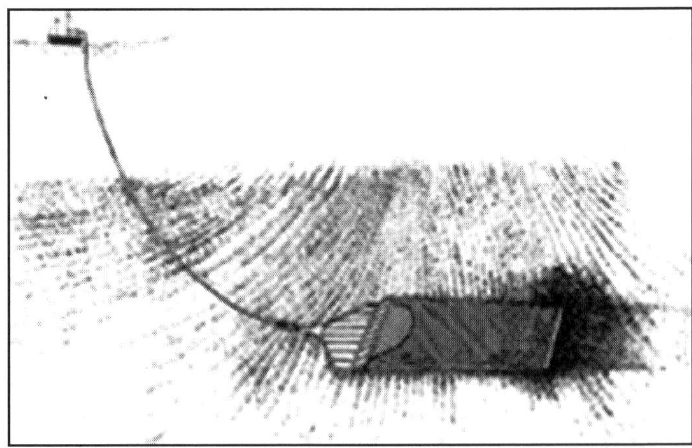

Fig. Action of a scallop and mussel dredge as it is dragged over a soft bottom seabed. The micro-relief of the seabed is smoothed as the dredge is towed over it. The gear in this figure is representative of scallop dredges used in many U.S. fisheries.

Offshore vessels (>30 m) slowly tow dredges as wide as 4.5 m across the seabed. The vessels have large hydraulic pumps that inject water into the sediment through flexible hoses connected to a manifold with multiple nozzles ahead of the blade of the dredge. The dredge is towed slowly so as to not exceed the liquefaction rate. When operated correctly, these dredges are highly efficient, taking as much as 90 percent of the clams in their path.

Management of effort in dredge fisheries is generally achieved with time and area closures and with restrictions on the size (blade width and weight) of dredges, the number of dredges, and the size and horsepower of the towing vessels.

FISHING DREDGE

Scallop dredges used by NOAA

A fishing dredge, also known as a scallop dredge or oyster dredge, is a kind of dredge which is towed along the bottom of the sea by a fishing boat in order to collect a targeted edible bottom-dwelling species. The gear is used to fish for scallops, oysters and other species of clams, crabs, and sea cucumber. The dredge is then winched up into the boat and emptied. Dredges are also used in connection with the work of the naturalist in marine biology, notably on the Challenger Expedition.

CONSTRUCTION

The dredge is usually constructed from a heavy steel frame in the form of a scoop. The frame is covered with chain mesh which is open on the front side, which is towed. The chain mesh functions as a net.

Dredges may or may not have teeth along the bottom bar of the frame. In Europe, early dredges had teeth, called tynes, at the bottom. These teeth raked or ploughed the sand and mud, digging up buried clams. This design was improved by using spring-loaded teeth that reduced bottom snagging, and so could be used over rough ground. The New Bedford (USA) dredge does not have teeth. Dredge nets have a coarse mesh in order to let organisms smaller than the target species through. The net catches the larger organisms: in the case of scallop dredging that includes the scallops' predators, such as whelks, starfish and octopus. In some cases several dredges are attached to a wheeled rigid axle in groups of three or four. A number of these dredges can be towed from a heavy spreading bar, usually one from each side of the vessel. The length of the bar and number of dredges towed depends on the power of the vessel and the room on the side deck for working the dredges. The number might be three on each side on a small 10-metre (33 ft) boat up to 20 on each side for a 30-metre (98 ft) vessel with 1500 hp. The great weight and strength of the gear can disturb the ground it is towed over, overturning rocks and dislodging and crushing organisms in its path.

Scallop dredging tends to result in scallops containing grit, and can damage the seabed if done carelessly. So these days scallop dredging is often replaced by scuba diving.Like a better mouse trap, there is still a challenge for inventors to produce a kinder and more gentle scallop dredge.

DREDGES – THE BOTTOM SCRAPERS

Dredges consist of a metal mouth frame with an attached collection bag. The collection bag is made of metal rings, metal bars, and/or mesh webbing. Fishing vessels drag the dredge along the sea floor to scoop up bottom-living species.

There are two types of dredges:

- Scraping dredges have teeth or a cutting bar that dig into the bottom sediment to collect animals that live on or just under the sediment surface.
- Penetrating dredges (often called hydraulic dredges) shoot a jet of water into the seafloor to chase animals that live deep in the sediment out and into the collection bag.

Dredges used in offshore waters are typically 12-15 feet (3.5-4.5 meters) wide and may weigh 2,600 pounds (1,000 kg) or more. In shallow habitats, smaller dredges are used. Fishing vessels may drag one or more dredges at a time. Typical drag times range from 15 minutes to 1 hour.

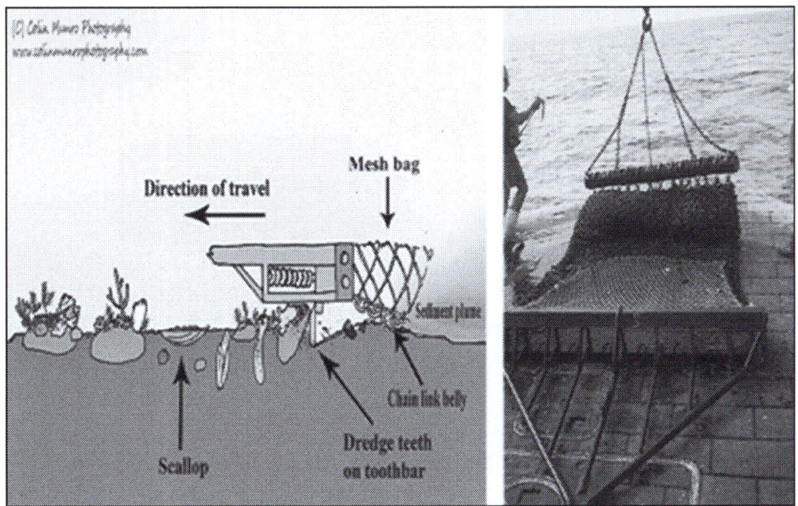

Left: Illustration showing the key parts of a scraping dredge, how it works, and how it can affect bottom habitats and species. Credit: Colin Munro Photography, colinmunrophotography.com. Right: Scraping dredge used to collect Atlantic sea scallops. Credit: NOAA.

WHAT DOES A DREDGE CATCH?

Dredges are primarily used to catch clams, scallops, oysters, or mussels – collectively referred to as 'bivalves' because they have two valves or shell parts. Generally, scraping dredges are used to collect scallops, oysters, and mussels. Penetrating or hydraulic dredges are used to catch some clam species (e.g., Atlantic surf clam, Ocean quahog). Dredges are also sometimes used to catch crabs, sea urchins, sea cucumbers, and conch.

Left: A hydraulic clam dredge with water coming from jets as it is lifted out of the water. Right: Catch from a clam dredge. Credit: NOAA.

HOW DO DREDGES AFFECT THE OCEAN?

Dredges rival bottom trawls for the title "most damaging gear to bottom ocean habitats". As the dredge is dragged along the sea floor, it leaves large scars, stirs up the sediment (which can reduce water quality), plows over or uproots structural features, like rocks, sponges, and seagrass, and may bury, dislodge, or kill bottom-living species such as worms, snails, and crabs. Dredges that have long teeth (like a rake) are particularly damaging because they dig deep into the sediment.

Penetrating or hydraulic dredges, which shoot jets of water into the sea floor, also cause a massive disturbance to bottom habitats. They liquefy the sediments and leave a trench in the sea floor that is typically the width of the dredge (12-15 ft) and up to 1 ft deep. These trenches may last for several months to more than a year.

Dredging can negatively affect the habitat of the very species fishermen are trying to catch and the habitats of other ocean animals too. For instance, dredging in seagrass areas greatly reduces the abundance of seagrass. This in turn negatively affects populations of bay scallops, mussels, crabs, lobsters, and fish that rely on seagrass for critical nursery habitat and protection from predators. It may take several years for the seagrass to re-grow. Dredging over oyster beds [a large aggregation of oysters] can reduce the height and alter the structure of the beds. This may make the beds less suitable for oyster growth and survival. And it can reduce the amount of suitable habitat for small fish, shrimp or other species that live in the small spaces between the oysters. In a variety of habitat types, scientists have found that dredged areas have a reduced number and variety of bottom-living species compared to similar un-dredged areas.

Some dredge fisheries also catch significant amounts of non-target species or 'bycatch'. This is often a concern in scallop fisheries. The U.S. Atlantic sea scallop fishery catches several bottom-living fish species such as monkfish, flounders, and skates, undersized scallops, and sometimes loggerhead sea turtles.

The bycatch can be 10-30% of the total catch. The UK Scottish Scallop fishery catches a significant amount of bycatch that includes starfish, sea urchins, whelks, undersized scallops, and other shellfish. Fishermen discard much of the bycatch – some dead and some alive – back to sea. This can attract predators and it is likely that species damaged during capture are quickly gobbled up.

Other dredge fisheries, however, such as those for oysters, clams, or mussels are fairly selective and capture low amounts of non-target species.

WHAT CAN BE DONE TO LESSEN THE NEGATIVE EFFECTS OF THIS FISHING GEAR?

To reduce the negative effects dredge fishing has on bottom ocean habitats, fishery managers can restrict the size and number of dredges that are used. In shallow water fisheries, lighter-weight dredges are often used, which helps reduce the amount of penetration into the sediment. But the best way to reduce negative effects to the habitat is to close off large areas to dredging. Fishery managers have done this in the Alaska Weathervane Scallop fishery. Fishing for Weathervane Scallops takes place over a very small proportion of the available habitat. This helps ensure a sufficient amount of habitat remains healthy and robust for scallops and other species. Because of the significant measures to protect habitat in this fishery, the Alaska Weathervane Scallop is rated "green"! Fishery managers of the U.S. Atlantic Sea Scallop fishery have also closed some areas to scallop dredging. This has benefited both sea scallops and ocean habitats.

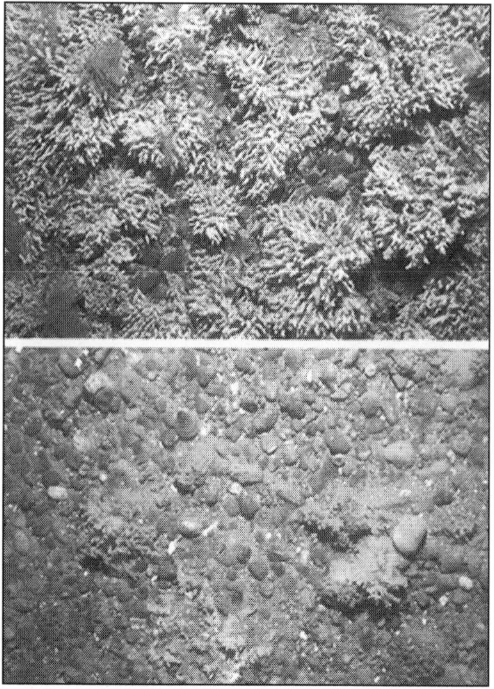

The top photo shows a normal sea floor community on Georges Bank, while the bottom photo shows a similar area after dredges have been used to catch scallops. Credit: Dann Blackwood and Page Valentine, USGS.

To reduce the amount of bycatch or non-target catch in dredge fisheries, fishery managers may require a minimum ring size or a minimum spacing between bars on the collection bag to allow small animals to pass through and avoid capture. [However, this does not necessarily prevent those small animals from getting run over and crushed by the dredge gear]. In some cases, fishery managers have restricted fishing in areas where high bycatch is known to occur and placed limits on the amount of allowed bycatch.

In the U.S. Atlantic sea scallop fishery, fishermen are required to utilize turtle deflector dredges in certain areas and during certain times when the bycatch of sea turtles is a concern. This modified dredge is designed to deflect sea turtles over the frame and collection bag to prevent them from getting caught or injured.

THE PATH FORWARD

It's clear that dredges can cause severe damage to ocean bottom habitats and species. So we must limit the amount of habitat that is affected. It is imperative that fishery managers close off large areas to dredge fishing to allow destroyed habitats to begin to recover and to protect any remaining undisturbed areas.

Additionally, in fisheries where dredging takes place in sensitive habitats, like seagrass areas, oyster beds, or mussel beds, the fishery should consider switching to less damaging fishing methods. We will cover some of these less damaging methods, which include hand rakes, hand tongs, or diving, later in this series. Reducing damage to these valuable habitats would greatly benefit our coastal ocean ecosystems, since they provide essential habitat for many ocean animals, help limit coastal erosion, and in the case of oysters and mussels, filter the water. And this would benefit the very shellfish populations fishermen want to catch!

TRAWL GEAR

As fishermen sought to increase efficiency, their vessels became larger and more powerful. This facilitated the evolution of dredges into beam trawls and otter trawls.

BEAM TRAWLS

With the beam trawl, the steel frame used previously in dredges became larger and lighter and the bag became larger and funnel-shaped to concentrate the catch in a cylindrical webbing section, called the codend. The first beam trawls were towed by sailing vessels, but today large beam trawls, with mouth openings of 15–20 m, are towed from both sides of high horsepower trawlers. Beam trawls do not open or sweep areas that are as large as those covered by otter trawls, but are used in selected flatfish fisheries because the beam maintains a constant opening and allows for the attachment of many heavy tickler chains to the frame.

The chains disturb the seabed ahead of the sweep of the net to stimulate an escape response in flatfish that have burrowed into the sediment. The gear is towed quickly, up to 2.5 m/s so escape is unlikely and catch efficiency is high. The frame of the beam trawl consists of an elevated pipe or beam and shoes that tend the seabed at each side. The beam is the site of attachment both for the upper section of the net and for the bridle, which distributes the towing force. The vertical opening of the net is restricted to the height of the beam above the seabed. The sweep of the net is attached to the shoes of the beam, as are tickler chains, if they are used.

Beam trawls are used in flatfish fisheries in northern Europe to catch sole and other groundfish species, and they have been used in experimental fisheries for monkfish, shrimp, and other demersal species in the United States. A modification of the traditional beam trawl is the plumbstaff trawl used by small gillnet vessels (about 10 m) in Alaska. Beam trawls are sometimes used in scientific sampling because they have fixed mouth openings that allow for accurate determination of sampling areas.

In Florida, roller frame trawls are used to harvest bait shrimp. They are similar to beam trawls in that there is a rectangular, rigid frame that holds the net open. The bottom of the frame supports freely turning, slotted rollers that allow the trawl to move over rough bottoms with less scraping of the seafloor than is typical with otter trawls (Barnette, 2001).

OTTER TRAWLS

Otter trawls developed as fishermen sought to increase the horizontal opening of the trawl mouth, but without the cumbersome rigid beam. In the late 1880s, Musgrave invented the otter board, a water-plane device that when used in pairs, each towed from a separate wire, opens the net mouth horizontally and holds the net on the bottom. Initially, all otter boards were connected to the wing ends of the trawl, as they are today in the shrimp trawl fishery. In the 1930s, the Dan Leno gear was developed in France by Vigarnon and Dahl. This gear allowed the otter boards (doors) to be separated from the trawl wing ends using cables or "ground gear." This technology increased the effective area swept from the distance between the net wings to the distance between the doors. The ground gear can be as long as 200 m, thus increasing the area swept by the trawl as much as three-fold.

The funnel-shaped bottom trawl net consists of upper and lower sections joined at seams (a twoseam net), called gores. Some fourseam bottom trawls also have side panels to increase vertical opening. The mouth of the trawl net consists of jib and wing sections in the upper and lower panels. A "square" section forms a roof over the net mouth. The body of the trawl net includes belly sections, leading to the codend, where the catch is collected. The webbing is attached to a rope frame consisting of a headrope along the leading edge of the upper panel, and a footrope along the leading edge of the lower panel. The sweep, which tends bottom as the net is towed, is attached to the footrope.

The headrope is equipped with floats that open the net mouth vertically. The headrope and footrope or sweep are attached to bridles (also called legs) at the wing ends that lead to the ground wires and the trawl doors.

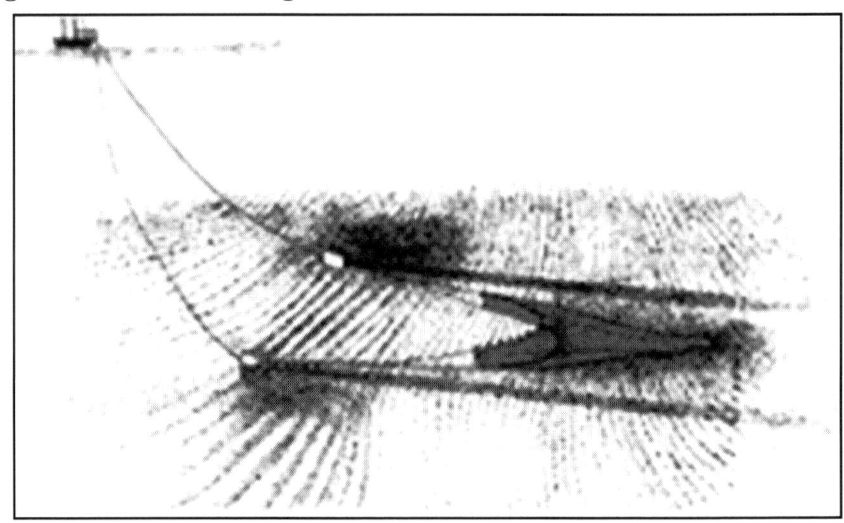

Fig. Action of bottom trawl as it is dragged over a soft-bottom seabed. The otter boards disturb the seabed creating visible tracks and suspending clouds of sediment in their wake.

The trawl doors, ground gear, and sweep all disturb the seabed as the gear is towed on the bottom. The magnitude of the interaction is a function of the design and operation of the specific trawl component and the nature of the seabed. Otter doors range in design from the traditional flat, rectangular boards that create great turbulence in their wake to the modern slotted or foil board that reduces turbulent drag by maximizing hydrodynamic efficiency. Trawl sweeps range in design from simple looped chain sweeps used on smooth, soft bottoms to large roller and rockhopper sweeps used on hard, rough bottoms. The development of this more sophisticated trawl sweep technology (rollers and rockhoppers)—combined with sonar technology and the precise navigation afforded by GPS—allows fishermen to trawl areas of the seabed previously considered too rough to fish with mobile gear. However, sonar and GPS also have made it easier for fishermen to avoid areas with rough seabed and thus minimize damage and loss of gear.

There are bottom trawl fisheries for demersal species on all U.S. coasts. In the northeast, 15–50 m vessels fish in waters that are 10–400 m deep. Small mesh nets catch northern shrimp, silver hake, butterfish, and squid. Large mesh trawls catch cod, haddock, flounders, and other large species. Those trawls typically are rigged with long ground wires that create sand clouds on the seabed, herding the fish into the trawl mouth. In the southeast and Gulf Coast areas, small mesh trawls catch shrimp. Because shrimp cannot be herded, shrimp trawl nets are usually connected directly to the trawl doors. Southern shrimp trawl vessels tow two to four trawls from large booms extended from each side of the vessel. On the West Coast, stern trawlers catch shrimp, flatfish, and various gadoid and rockfish species. Factory trawlers or catcher processors, from 50–100 m long, catch, process, and freeze their products onboard.

Pair bottom trawling, once used to catch groundfish in New England waters, is undertaken by two vessels towing a single net. The separation of the towing vessels is used to open the net mouth horizontally. Pair bottom trawls are generally no larger than nets towed by single vessels. The advantage of pair bottom trawling is that considerably more ground gear can be used so as to increase the area swept, due to the reduction in drag resulting from the absence of trawl doors. Use of this gear is now prohibited in the New England groundfish fisheries.

TRAWLING

Trawling is a method of fishing that involves pulling a fishing net through the water behind one or more boats. The net that is used for trawling is called a trawl. The boats that are used for trawling are called trawlers or draggers. Trawlers vary in size from small open boats with only 30 hp engines to large factory trawlers with over 10,000 hp. Trawling can be carried out by one trawler or by two trawlers fishing cooperatively (pair trawling).

Trawling can be contrasted with trolling, where baited fishing lines instead of trawls are drawn through the water. Trolling is used both forrecreational and commercial fishing whereas trawling is used mainly for commercial fishing. Trawling is also commonly used as a scientific sampling, or survey, method.

BOTTOM VERSUS MIDWATER TRAWLING

Trawling can be divided into bottom trawling and midwater trawling, depending on how high the trawl (net) is in the water column. Bottom trawling is towing the trawl along (benthic trawling) or close to (demersal trawling) the sea floor. Midwater trawling is towing the trawl through free water above the bottom of the ocean or benthic zone.

Midwater trawling is also known as pelagic trawling. Midwater trawling catches pelagic fish such as anchovies, shrimp, tuna andmackerel, whereas bottom trawling targets both bottom-living fish (groundfish) and semi-pelagic fish such as cod, squid, halibut androckfish.

The gear itself can vary a great deal. Pelagic trawls are typically much larger than bottom trawls, with very large mesh openings in the net, little or no ground gear, and little or no chaffing gear. Additionally, pelagic trawl doors have different shapes than bottom trawl doors, although doors that can be used with both nets do exist.

NET STRUCTURE

When two boats are used (pair trawling), the horizontal spread of the net is provided by the boats, with one or in the case of pelagic trawling two warps attached to each boat. However, single-boat trawling is more common. Here, the horizontal spread of the net is provided by trawl doors (also known as "otter boards"). Trawl doors are available in various sizes and shapes and may be specialized to keep in contact with the sea bottom (bottom trawling) or to remain elevated in the water. In all cases, doors essentially act as wings, using a hydrodynamic shape to provide horizontal spread. As with all wings, the towing vessel must go at a certain speed for the doors to remain standing and functional. This speed varies, but is generally in the range of 2.5–4.0 knots. The vertical opening of a trawl net is created using flotation on the upper edge ("floatline") and weight on the lower edge ("footrope") of the net mouth. The configuration of the footrope varies based on the expected bottom shape. The more uneven the bottom, the more robust the footrope configuration must be to prevent net damage. This is used to catch shrimp, shellfish, cod, scallops and many others. Trawls are funnel-shaped nets that have a closed-off tail where the fish are collected and is open on the top end as the mouth. Trawl nets can also be modified, such as changing mesh size, to help with marine research of ocean bottoms.

ENVIRONMENTAL EFFECTS

Although trawling today is heavily regulated in some nations, it remains the target of many protests by environmentalists. Environmental concerns related to trawling refer to two areas: the lack of selectivity and the physical damage which the trawl does to the seabed.

Selectivity

Since the practice of trawling started (around the 15th century), there have been concerns over trawling's lack of selectivity. Trawls may be non-selective, sweeping up both marketable and undesirable fish and fish of both legal and illegal size. Any part of the catch which cannot be used is considered by-catch, some of which is killed accidentally by the trawling process. By-catch commonly includes valued species such as dolphins, sea turtles, and sharks, and may also include sublegal or immature individuals of the targeted species.

Many studies have documented large volumes of by-catch that are discarded. For example, researchers conducting a three-year study in theClarence River found that an estimated 177 tons of by-catch (including 77 different species) were discarded each year. Size selectivity is controlled by the mesh size of the "cod-end"—the part of the trawl where fish are retained. Fishermen complain that mesh sizes which allow undersized fish to escape also allows some legally–catchable fish to escape as well. There are a number of "fixes", such as tying a rope around the "cod-end" to prevent the mesh from opening fully, which have been developed to work around technical regulation of size selectivity. One problem is when the mesh gets pulled into narrow diamond shapes (rhombuses) instead of squares.

The capture of undesirable species is a recognized problem with all fishing methods and unites environmentalists, who do not want to see fish killed needlessly, and fishermen, who do not want to waste their time sorting marketable fish from their catch. A number of methods to minimize this have been developed for use in trawling. Bycatch reduction grids or square mesh panels of net can be fitted to parts of the trawl, allowing certain species to escape while retaining others. Studies have suggested that shrimp trawling is responsible for the highest rate of by-catch.

Environmental Damage

Trawling is controversial because of its environmental impacts. Because bottom trawling involves towing heavy fishing gear over the seabed, it can cause large-scale destruction on the ocean bottom, including coral shattering, damage to habitats and removal of seaweed. The primary sources of impact are the doors, which can weigh several tonnes and create furrows if dragged along the bottom, and the footrope configuration, which usually remains in contact with the bottom across the entire lower edge of the net. Depending on the configuration, the footrope may turn over large rocks or boulders, possibly dragging them along with the net, disturb or damagesessile organisms or rework and re-suspend bottom sediments. These impacts result in decreases in species diversity and ecological changes towards more opportunistic organisms. The destruction has been likened to clear-cutting in forests.

The primary dispute over trawling concerns the magnitude and duration of these impacts. Opponents argue that they are widespread, intense and long-lasting. Defenders maintain that impact is mostly limited and of low intensity compared to natural events. However, most areas with significant natural sea bottom disturbance events are in relatively shallow water. In mid to deep waters, bottoms trawlers are the only significant area-wide events.

Bottom trawling on soft bottoms also stirs up bottom sediments and loading suspended solids into the water column. One bottom trawler can put more than 10 times the amount of suspended solids pollution per hour into the water column than all the suspended solids pollution from all the sewerage, industrial, river and dredge disposal operations in Southern California combined. These turbidity plumes can be seen on Google Earth in areas where they have high resolution offshore photos. When the turbidity plumes from bottom trawlers are below a thermocline, the surface may not be impacted, but less visible impacts can still occur, such as persistent organic pollutant transfer into the pelagic food chain.

As a result of these processes, a vast array of species are threatened around the world. In particular, trawling can directly kill coral reefs by breaking them up and burying them in sediments. In addition, trawling can kill corals indirectly by wounding coral tissue, leaving the reefs vulnerable to infection. The net effect of fishing practices on global coral reef populations is suggested by many scientists to be alarmingly high. Published research has shown that benthic trawling destroys the cold-water coral Lophelia pertusa, an important habitat for many deep-sea organisms.

Midwater (pelagic) trawling is a much "cleaner" method of fishing, in that the catch usually consists of just one species and does not physically damage the sea bottom. However, environmental groups have raised concerns that this fishing practice may be responsible for significant volumes of by-catch, particularly cetaceans (dolphins, porpoises, and whales).

REGULATION

In light of the environmental concerns surrounding trawling, many governments have debated policies that would regulate the practice.

OTHER USES OF THE WORD "TRAWL"

The noun "trawl" has many possibly confusing meanings in commercial fisheries. For example, two or more lobster pots that are fished together may be referred to as a trawl. In some older usages "trawling" meant "long-line fishing"; that usage occurs in Rudyard Kipling's book Captains Courageous. (This use is perhaps confused with trolling, where a baited line is trailed behind a boat. Troll also has several meanings.) The word "trawling" has come to be used in a number of non-fishing contexts, usually meaning indiscriminate collection with the intent of picking out the useful bits. For instance, in law enforcement it may refer to collecting large volumes of telephone call records hoping to find calls made by suspects. It also occurs frequently in reference to research methods, where it means searching through written sources for relevant information.

TRAWLS – BULLDOZERS OF THE OCEAN

Trawls are enormous, cone-shaped nets that are towed by one or two boats. As the net is towed, it herds and captures thousands of fish and other creatures. The net is wide at the mouth and then narrows to a bag or 'cod-end', where the fish are trapped.

When towed by a single vessel, heavy wood or steel doors on each side of the net (called "otter boards") or a solid beam, hold the mouth of the net open. When towed by two vessels, the distance between the vessels keeps the mouth of the net open. The spread of the trawl net can be up to 330 feet (100 meters) wide and 40 feet (12 meters) high. Picture it this way – a net as wide as a football field and higher than a three-story house.

Trawl fishing can take place at shallow depths to extreme depths:

- When the net is towed along the bottom of the ocean floor, it is called a "bottom trawl".
- When towed off the bottom, it is called a "mid-water trawl" or "pelagic trawl".

The size of the net openings (or "mesh size"), tow speed and tow duration will vary depending on the species fishermen are targeting.

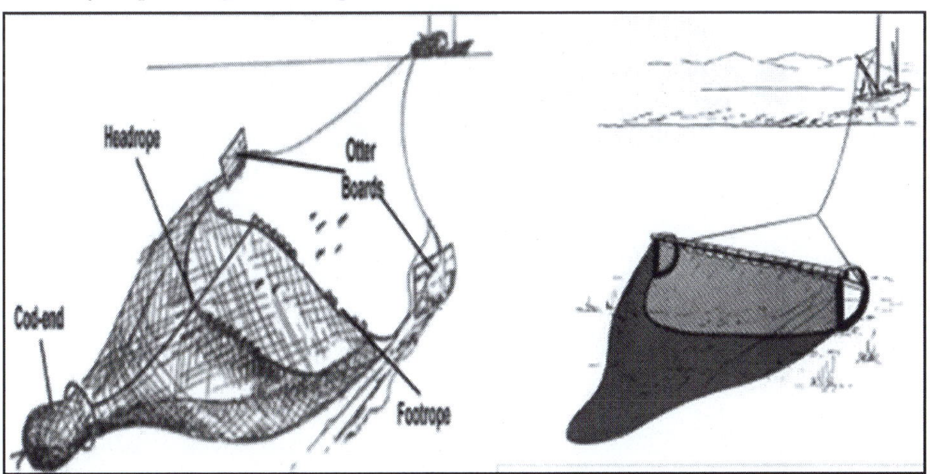

Left: A trawl held open by otter boards. The force of water pressure against the boards as it is towed keeps the net open. The bottom of the net, 'the footrope' has weights, while the top of the net, 'the headrope' has floats.

WHAT DOES A TRAWL CATCH?

Bottom trawls are used to catch fish that live on or near the bottom of the ocean floor, such as cod, haddock, flounder, sole, rockfish, and orange roughy, as well as shrimp, octopus, and squid. Mid-water trawls are used to catch species that tend to gather in large groups or "schools" in the top or mid depths of the ocean, like Alaskan pollock, mackerel, herring, and sardines.

HOW DO TRAWLS AFFECT THE OCEAN?

Trawls are highly indiscriminate, capturing any and all species in their path. They catch high amounts of species that fishermen are not trying to catch, termed 'bycatch'.The unintended catch or bycatch may include many species of fish, invertebrates (such as crabs, scallops, starfish or corals), sharks, skates and rays, endangered sea turtles, and sometimes whales and dolphins. Fishermen often throw much of this unintended catch back to sea dead or dying. This is a giant waste of marine life! Shrimp trawls in particular catch very high amounts of bycatch– often 2-10 times the amount of shrimp caught.

Fig. Catch from a trawl net. Photo by Carl Safina.

Because mid-water trawls typically target specific schools of fish, they tend to catch lower amounts of bycatch compared to bottom trawls. But mid-water trawls can still sometimes have a devastating effect on species of fish that are not sought after by the fishery or on ocean wildlife, like sea turtles, dolphins, and whales. For example, U.S. mid-water trawlers that target Atlantic herring and mackerel often catch significant numbers of depleted river herring [an important Atlantic fish that migrates between river waters and ocean waters]. A single trawl tow may catch as many as 250,000 river herring. And the trawls sometimes catch pilot whales, white-sided dolphins, and common dolphins.

Bottom trawls also cause significant damage to habitats on the bottom of the ocean.The trawl nets and their associated parts are very heavy. As these nets are dragged along the sea floor, they stir up the sediment, crush bottom-dwelling species like crabs and snails, destroy structural organisms like coral, sponges, or seagrass, and displace rocks and boulders.

Because of this damage, trawled areas do not make good habitats for ocean animals. Scientists have found that areas that are frequently bottom trawled often have a reduced number and variety of bottom-living invertebrates (such as worms, snails, clams, or lobster) and fish.Bottom trawling turns physically diverse environments, teeming with life into barren deserts. Essentially, it is equivalent to clear-cutting the rainforest, especially when trawling over coral reefs.

It should be noted that while mid-water trawls are towed off the bottom, they can still sometimes contact the ocean bottom and cause damage. For instance, mid-water trawls that target Alaska pollock and Atlantic herring or mackerel are known to occasionally contact the ocean bottom.

WHAT CAN BE DONE TO LESSEN THE NEGATIVE EFFECTS OF THIS FISHING GEAR?

Turtle excluder devices and bycatch reduction devices are two common types of gear modifications that are used in trawl fisheries to reduce bycatch. The idea of these devices is to provide a way for large or unwanted animals to escape, while still allowing for the capture of targeted species.

Turtle excluder devices are fitted into the neck of the trawl and consist of a grid of bars plus an escape opening/door at either the top or bottom of the net. When a large animal like a turtle hits the bars, it is directed to the escape opening and is able to swim free.

Small animals like shrimp and fish, however, pass through the bars to the bag of the net. I know this is hard to visualize, so to see exactly how this works watch this short video.

Bycatch reduction devices work similar to turtle excluder devices. For trawl fisheries targeting shrimp, these devices are designed to exclude fish, while still allowing shrimp to pass into the bag of the net. In trawl fisheries for fish, bycatch reduction devices are sometimes used to prevent the capture of a particular depleted species. There are various types of bycatch reduction devices– below are a few examples.

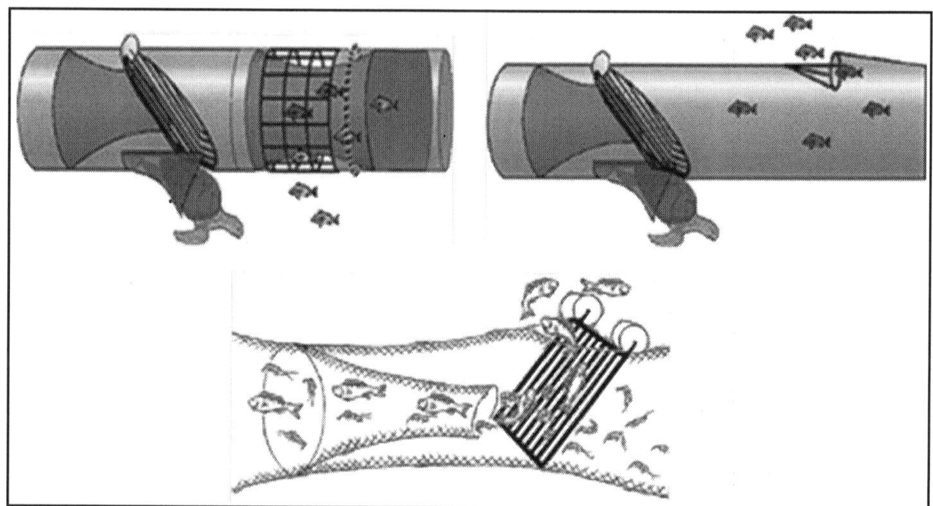

Examples of various types of bycatch reduction devices. Top left: Funnel bycatch reduction device – Uses an inner small mesh funnel to provide a passage for shrimp to the bag of the net while a large mesh outer layer provides a means for large fish to escape. The design creates an area of reduced water flow so the fish can easily swim out the large mesh openings. Top right: The Fisheye – A cone-shaped rigid frame forms on opening in the net for fish to escape. Bottom: Sorting grate – Shrimp pass through the grate, while fish that are not large enough to pass through are pushed out through the escape opening.

When used properly, turtle excluder devices and bycatch reduction devices can significantly reduce the deaths of sea turtles and other unintended species. Although they don't reduce bycatch completely, these devices are an important step in the right direction.

In the U.S., the majority of shrimp trawl fisheries are required to install turtle excluder devices and/or bycatch reduction devices in their nets. We have also helped encourage the use of turtle excluder devices in other shrimp fisheries around the world.

In fact, the U.S. has a law that prohibits importing shrimp from countries that do not require the use of turtle excluder devices when fishing in areas where turtles live. Currently, around 40 countries are certified to export shrimp to the U.S.

However, there are many trawl fisheries that are not using turtle and bycatch excluder devices, and fishermen compliance with these excluder device regulations remains an issue in many cases – even in some U.S. fisheries (e.g. Louisiana shrimp fishery). There are also gear modifications or technologies that can help limit the effects trawls have on bottom ocean habitats. For instance, fishermen can add disks to the bottom of the trawl net to elevate it slightly off the seafloor, reducing the amount of contact the gear has with the bottom. This has been implemented in the Alaska flatfish trawl fishery and has been shown to significantly reduce damage to bottom habitats.

Of course, the best way to protect ocean habitats is to limit the areas where trawling is allowed to occur. It is particularly important that fishery managers prohibit trawling in habitats that are especially sensitive to trawl damage, such as coral reefs, sponges, and seagrass beds and in important fish habitats.

An un-trawled coral reef area in the Oculina Banks off the coast of Florida is pictured on the left, while a trawled area is pictured on the right.

As we have learned more about the lasting effects that trawl fishing has on the ocean, several countries are banning bottom trawling in sensitive habitat areas and large areas off their coasts. The U.S. has banned bottom trawling in large areas off the U.S. Pacific Coast and in some sensitive habitats off the U.S. East Coast.

A few countries (e.g., Palau, Belize) have banned bottom trawling in all of their waters (kudos to them).

THE PATH FORWARD

Despite the destruction trawls cause, they remain one of the most common methods for catching fish and shrimp and many ocean areas remain vulnerable to trawl fishing.Across the globe, fishery managers must take measures to limit the effects of trawling before they turn vast swaths of the ocean into barren wastelands. And managers must take measures to recover places that trawling has destroyed.

We must ensure that turtle excluder devices and bycatch reduction devices are used and enforced in all trawl fisheries that catch vulnerable species. However, even with these excluder devices, trawl fisheries can still catch large numbers of unintended marine animals. Therefore when possible, managers and fishermen should look to fishing with more selective fishing gear. For example, some shrimp fisheries are using pots instead of trawls (to be covered later in the blog series—so stay tuned).

To protect ocean habitats, fishery managers must close off more areas to trawling and ensure that bottom trawling is not allowed in any new areas. (We have trawled enough of the ocean already!) We should prohibit bottom trawling (and mid-water trawls that contact the bottom) in the deep-sea and in any areas with coral reefs, sponges, seamounts, or other sensitive habitat features. But we must also protect large portions of all habitat types from this destructive practice to make sure that all ocean species have healthy places to live.

8

Passive Fishing Gears

Passive gears are in general the most ancient type of fishing gears. These gears are most suitable for small scale-fishing and are, therefore, often the gear types used in artisanal fisheries. Some passive fishing gears are often referred to as "stationary" fishing gears. Stationary gears are those anchored to the seabed and they constitute a large group of the passive gears. However, some moving gears such as drift nets may also be classified as passive gears, as fish capture by these gears also depend on movement of the target species towards the gear.

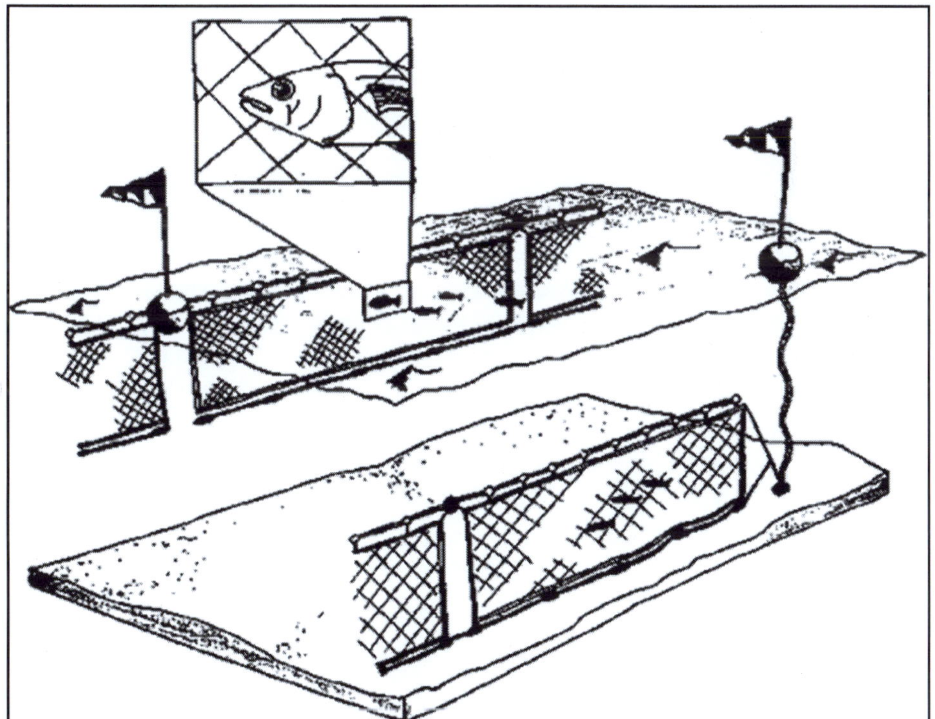

Fig. Catching principle (expanded view) and construction of gillnets, pelagic drift net (top) and bottom set (lower).

NETS

A fishing net or fishnet is a net used for fishing. Nets are devices made from fibers woven in a grid-like structure. Fishing nets are usually meshes formed by knotting a relatively thin thread. Early nets were woven from grasses, flaxes and other fibrous plant material. Later cotton was used. Modern nets are usually made of artificial polyamides like nylon, although nets of organic polyamides such as wool or silk thread were common until recently and are still used.

HISTORY

Fishing nets have been used widely in the past, including by stone age societies. The oldest known fishing net is the net of Antrea, found with other fishing equipment in the Karelian town of Antrea. The net was made from willow, and dates back to 8300 BC. The remnants of another fishing net dates back to the late Mesolithic, and were found together with sinkers at the bottom of a former sea. Some of the oldest rock carvings at Alta (4200–500 BC) have mysterious images, including intricate patterns of horizontal and vertical lines sometimes explained as fishing nets. American Native Indians on the Columbia River wove seine nets from spruce root fibers or wild grass, again using stones as weights. For floats they used sticks made of cedar which moved in a way which frightened the fish and helped keep them together. With the help of large canoes, pre-European Maori deployed seine nets which could be over one thousand metres long. The nets were woven from green flax, with stone weights and light wood or gourd floats, and could require hundreds of men to haul.

Fishing nets are well documented in antiquity. They appear in Egyptian tomb paintings from 3000 BC. In ancient Greek literature, Ovidmakes many references to fishing nets, including the use of cork floats and lead weights. Pictorial evidence of Roman fishing comes from mosaics which show nets. In a parody of fishing, a type of gladiator called retiarius was armed with a trident and a cast net.

He would fight against a secutor or the murmillo, who carried a short sword and a helmet with the image of a fish on the front. Between 177 and 180 the Greek author Oppian wrote the Halieutica, a didactic poem about fishing. He described various means of fishing including the use of nets cast from boats, scoop nets held open by a hoop, and various traps "which work while their masters sleep". Here is Oppian's description of fishing with a "motionless" net:

The fishers set up very light nets of buoyant flax and wheel in a circle round about while they violently strike the surface of the sea with their oars and make a din with sweeping blow of poles. At the flashing of the swift oars and the noise the fish bound in terror and rush into the bosom of the net which stands at rest, thinking it to be a shelter: foolish fishes which, frightened by a noise, enter the gates of doom. Then the fishers on either side hasten with the ropes to draw the net ashore.

In Norse mythology the sea giantess Rán uses a fishing net to trap lost sailors. References to fishing nets can also be found in the New Testament. Jesus Christ was reputedly a master in the use of fishing nets. The tough, fibrous inner bark of the pawpaw was used by Native Americans and settlers in the Midwest for making ropes and fishing nets. The archaeological site at León Viejo (1524–1610) has fishing net artifacts including fragments of pottery used as weights for fishing nets.

Fishing nets have not evolved greatly, and many contemporary fishing nets would be recognized for what they are in Neolithic times. However, the fishing lines from which the nets are constructed have hugely evolved. Fossilised fragments of "probably two-ply laid rope of about 7 mm diameter" have been found in one of the caves at Lascaux, dated about 15,000 BC. Egyptian rope dates back to 4000 to 3500 BC and was generally made of water reed fibers. Other rope in antiquity was made from the fibers of date palms,flax, grass, papyrus, leather, or animal hair. Rope made of hemp fibres was in use in China from about 2800 BC.

CHINESE FISHING NETS

In India, Chinese fishing nets (Cheena vala) are fishing nets that are fixed land installations for fishing. While commonly known as "Chinese fishing nets" in India, the more formal name for such nets is "shore operated lift nets". Huge mechanical contrivances hold out horizontal nets of 20 m or more across. Each structure is at least 10 m high and comprises a cantilever with an outstretched net suspended over the sea and large stones suspended from ropes as counterweights at the other end. Each installation is operated by a team of up to six fishermen. While such nets are used throughout coastal southern China and Indochina, in India they are mostly found in the Indian cities of Kochi and Kollam, where they have become a tourist attraction. The Indian common name arises because they are unusual in India and different from usual fishing nets in India.

The system is sufficiently balanced that the weight of a man walking along the main beam is sufficient to cause the net to descend into the sea. The net is left for a short time, possibly just a few minutes, before it is raised by pulling on ropes.

The catch is usually modest: a few fish and crustaceans — these may be sold to passers by within minutes. Rocks, each 30 cm or so in diameter are suspended from ropes of different lengths. As the net is raised, some of the rocks one-by-one come to rest on a platform thereby keeping everything in balance.

Each installation has a limited operating depth. Consequently, an individual net cannot be continually operated in tidal waters. Different installations will be operated depending on the state of the tide. It was earlier thought that the nets might have been introduced by the Chinese explorer Zheng He. The Chinese fishing nets have become a very popular tourist attraction, their size and elegant construction is photogenic and the slow rhythm of their operation is quite hypnotic. In addition, catches can be purchased individually and need be taken only a short distance to a street entrepreneur who will cook it.

GILLNETS NET

The catching principle and construction of gillnets.

Catching Principle

The gillnet is named after its catching principle, as fish are usually caught by "gilling" - i.e. the fish is caught in one of the meshes of the gillnet, normally by the gill region (between the head and the body). Thus, fish capture by gillnets is based on fish encountering the gear during feeding or migratory movements. As fish may avoid the gillnet if they notice the gear, catches are normally best at low light levels or in areas with turbid water.

Construction

A gillnet consists basically of a "wall" or panel (e.g. 5 by 30 m) of meshes made from fine thread. The mesh panel is mounted with reinforcing ropes on all sides.

To obtain a vertical position of the net in the sea, floats and weights are fastened at regular intervals to the top rope (float line, cork line) and bottom rope (sinker line, lead line), respectively. The size of the meshes and hanging ratio (number of meshes per length of gillnet) are chosen to fit the desired target species and size.

Mesh size is commonly given as the length (in mm) either of a whole stretched mesh or the half-length (also called bar-length). Today, gillnets are almost exclusively made from synthetic fibres, normally nylon (polyamide) - either as multifilament thread or monofilament (gut). The latter is increasingly being used because of its low visibility and correspondingly higher catch efficiency. Multi-monofilament is also becoming more common.

Operation

Gillnets are most commonly operated as a stationary gear anchored to the bottom at either end, but may also be so-called drift-nets which float freely in the water. Stationary nets may be set on the seabed, at different depths in the water column or with the float line at the surface. Similarly, drift-nets may be operated with the float line at the surface or suspended from surface floats and corresponding float lines to the desired fishing depth in midwater.

Gillnets may be operated from vessels ranging from the smallest non-mechanised fishing boats to big, well equipped vessels capable of large-scale deep sea fishing. The gear used in small- and large-scale fishing is basically the same: the unit gillnet.

However, with increased vessel size a larger number of net units can be carried and operated per day. Single gillnets are then linked into long fleets of up to several hundred nets.

Gillnets can also be operated from shallow to large depths and can be used for fishing on rough bottom and at wrecks. One specific problem with gillnets is so-called "ghost fishing".

This refers to gillnets that are lost (most commonly after being stuck on a rough bottom) and continue to catch and kill fish over long periods of time. The Code of Conduct requires that the incidence of ghost fishing should be minimised.

Target Species

Gillnets are used to catch a large variety of fish species. In general, bottom gillnets are used for catching demersal species like cod, flatfish, croakers and snapper, while pelagic gillnets are used for species like tuna, mackerel, salmon, squid and herring.

TRAMMEL NETS

The catching principle and construction of trammel nets.

Catching Principle

In trammel nets, fish are caught by entanglement, which is facilitated by its special construction of three panels of nets attached on the same rope with a high degree of slackness.

Construction

At first glance, a trammel net may look like a gillnet. However, while the gillnet has a single panel of meshes, the trammel net has three - one middle panel of small meshes and two side panels of larger meshes.

When a fish comes in contact with the net, it will press the small mesh net through an adjacent larger mesh so that it is caught by entangling or "pouching".

Operation

Trammel nets are usually set and operated like bottom set gillnets, mainly in small-scale, near-shore fisheries.

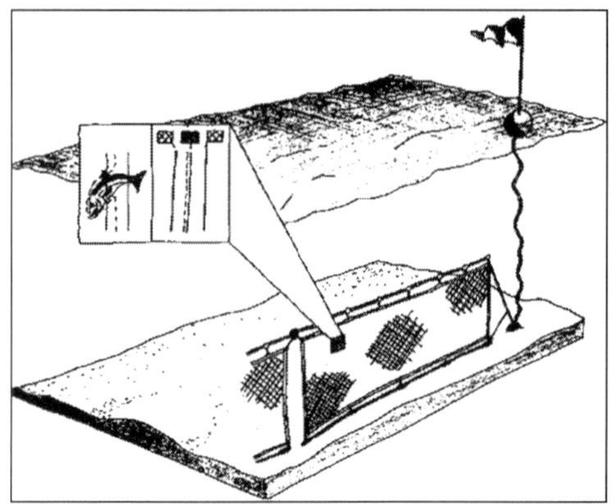

Fig. Catching principle (expanded view) and construction of trammel nets.

Target species

Trammel nets are used for catching a large variety of demersal fish.

HOOK AND LINE FISHING

Different fishing methods are based on the use of fish hooks; longlining, trolling and various forms of handlining such as jigging. The general catching principle of hook fishing is to attract the fish to the hook and entice the fish to bite and/or swallow the hook so that the fish becomes hooked and retained.

HOOK-AND-LINE

Fig. Hook and line.

Target

A wide variety of fish, squid and cuttlefish ranging from open ocean swimmers to bottom dwellers.

How it Works

Hook-and-line is a general term used for a range of fishing methods that employ short fishing lines with hooks in one form or another (as opposed to long-lines). It includes hand-lines, hand-reels, powered reels, rod/pole-and-line, drop lines, and troll lines, all using bait or lures in various ways to attract target species. Hand-lines use lines and baited hooks from a stationary or moving boat. Because hauling is slow, mechanised systems have been developed to allow more lines to be worked by a smaller crew. Pole-and-line, or 'bait-boat' fishing, attracts surface-schooling fish to the vessel, where they are driven into a 'feeding frenzy' by the throwing of live or dead bait into the water and the spraying of water onto the sea surface to simulate the escape of small prey. Lines are used to hook the fish, which are then pulled on board by manual or powered devices. This method is used to capture surface-schooling tuna such as skipjack and albacore. Trolling uses lines fitted with baits or lures that are trailed near the surface or at a certain depth by a vessel. Several lines are usually towed at the same time. These are commonly used for tuna and marlin. Jigging uses lures on a vertical line moved up and down (jigged) by hand or mechanically. When used at night, this method is highly efficient for catching oceanic squid, which are attracted by lights.

Advantages

Hook and line fishing is more selective than other types of fishing in terms of species and size, and provides high quality fish. The method can be used on spawning fish as they normally only bite after completion of spawning. Lines are set for a relatively short time so that any unwanted species can often be returned live to the sea.

PELAGIC LONG-LINES

Fig. Long-lines.

Target

A wide variety of pelagic species (those that spend most of their lives in the mid-water, with little contact with the seabed), but best known for tuna, swordfish and marlins.

How they Work

Long-lines consist of short lines (called snoods) carrying baited hooks, attached at regular intervals to a longer main line that is laid on the bottom or suspended horizontally with the help of surface floats. Main lines can be over 150 km long and can carry several thousand hooks.

Problems

Long-lines catch many endangered sharks, turtles, marine mammals and seabirds. Bycatch can be reduced in a variety of ways including circle hooks to prevent catching turtles, and setting deeper lines to reduce catches of turtles, sharks and marine mammals. Techniques such as setting lines quickly and at a greater depth, using bird scarers, and setting at night can reduce the number of seabirds that get caught on hooks and drown when diving for bait.

BOTTOM LONG-LINES

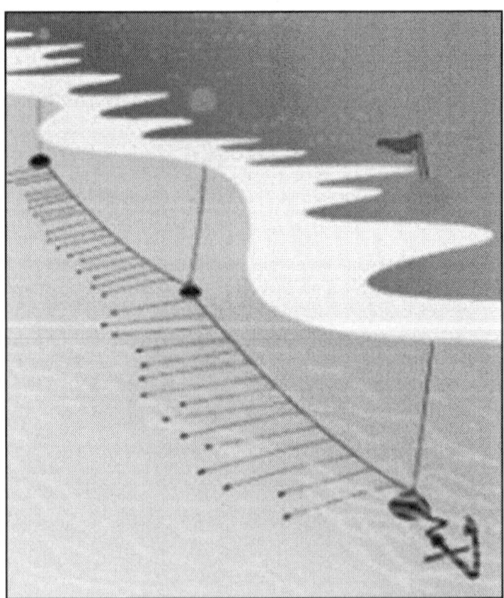

Fig. Bottom long-lines.

Target

A wide variety of benthic and demersal species (those that spend most of their lives on or near the seabed), including cods, haddock and flatfish.

How they work

Long-lines consist of short lines (called snoods) carrying hooks, attached at regular intervals to a longer main line that is laid on, or close to, the seabed. Main lines are up to 150 km long and can carry several thousand hooks.

Advantages

These have a much lower impact on the seabed than other bottom gears such as trawls.

Problems

Lost gear that gets snagged on the bottom may continue to 'ghost' fish for sometime, and there is bycatch of unwanted species.

PELAGIC GILLNETS

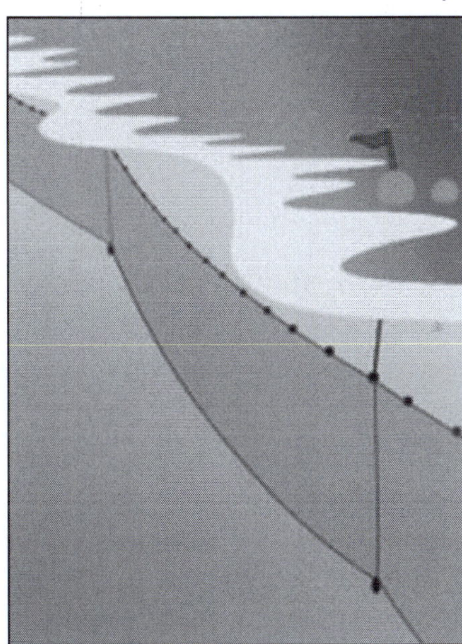

Fig. Pelagic gillnets.

Target

A wide variety of pelagic species (those that spend most of their lives in the mid-water, with little contact with the seabed), including salmon and herring.

How they Work

Pelagic gillnets or 'set nets' are fine-filament nets that are kept at or below the surface by numerous floats and weights and held in position by anchors. If a fish's head goes through the net but its body can't follow, it is 'gilled' or entangled in the netting when it tries to get out. Gillnets are used either alone or in large numbers placed in a row. Drifting gillnets or driftnets float freely with the current or with the boat to which they are attached, usually across the path of migrating fish schools.

Advantages

When set correctly and with the right mesh size, gillnets can be highly size selective: small fish can swim right through the net while larger fish will not get their heads stuck.

Problems

Poorly set gillnets or a bad choice of mesh size can result in higher levels of bycatch. Gillnets are associated with bycatch of marine mammals. Acoustic deterrents or 'pingers' attached to fishing nets may discourage some marine mammals from swimming too close and getting caught; however, trials of these have shown very mixed results for effectiveness, depending on the fishing area and the mammal species. Driftnets are associated with a high level of bycatch - large-scale pelagic driftnets are banned by a UN resolution. Lost or abandoned nets continue to entangle marine creatures - known as ghost netting.

BOTTOM GILLNETS

Fig. Bottom gillnets.

Target

A wide variety of benthic and demersal species (those that spend most of their lives on or near the seabed), including cods, haddock, flatfish and shrimp.

How they Work

Bottom gillnets or 'set nets' are fine-filament nets, the lower edge of which touch the seabed, and are held in place by numerous floats, weights and anchors. If a fish's head goes through the net but its body can't follow, it is 'gilled' or entangled in the netting when it tries to get out.

Gillnets are used either alone or in large numbers placed in a row. Trammel nets, a type of gillnet, are bottom-set nets made with three layers of netting, the two outer walls being of a larger mesh size than the loosely hung inner panel. The fish get entangled in the inner small-meshed panel after passing through one of the outer walls.

Advantages

When set correctly and with the right mesh size, these nets can be highly size selective: small fish can swim right throught the net while larger fish will not get their heads stuck. Bottom gillnets have a much lower impact on the seabed than other bottom gears such as trawls.

Problems

Poorly set gillnets or a bad choice of mesh size can result in higher levels of bycatch. Lost or abandoned nets continue to entangle marine creatures - known as ghost netting. Gillnets are associated with bycatch of marine mammals.

Acoustic deterrents or 'pingers' attached to fishing nets may discourage some marine mammals from swimming too close and getting caught; however, trials of these have shown very mixed results for effectiveness, depending on the fishing area and the mammal species.

HANDLINING AND TROLLING FISHING

The catching principle and construction of handlining.

Catching Principle

The fish is attracted to the hook by visual stimuli, either natural bait or more commonly in the form of artificial imitations of prey organisms like lures, jigs, rubber worms etc.

Construction

The gear is simple: a nylon monofilament is commonly used as line with one to several hooks at the end with bait or lures.

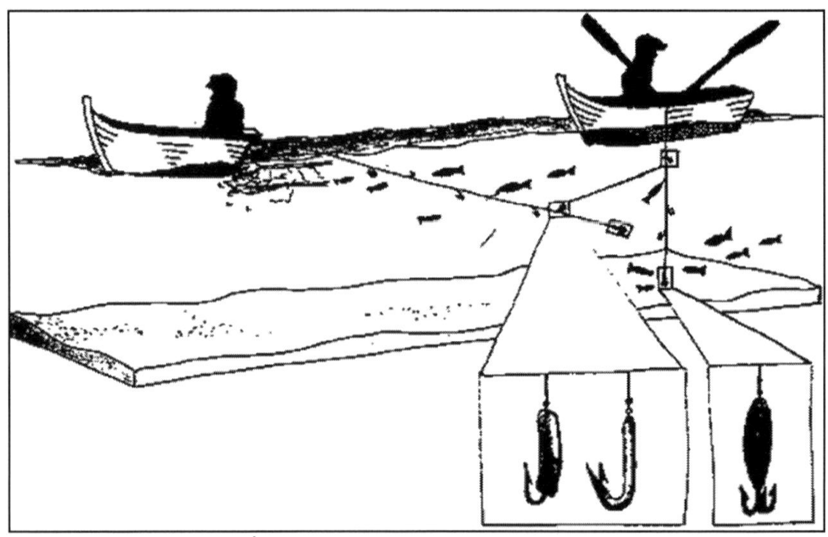

Fig. Catching principle and construction of trolling (left) and jigging (right). Expanded view: examples of lures and jig.

Operation

In handlining the fishing line is vertical and is operated from a drifting or anchored vessel. Handlining is also conducted from the shore, with and without the use of a pole.

From using only a single line, the operation can be scaled up by using several lines on larger vessels. In recent years jigging has become mechanised and automated by the development of jigging machines.

Hook and line can also be used in trolling where the fishing line is towed behind the moving vessel. Semi-automation has also been developed in trolling where power reels are often used for hauling the lines. Trolling is considered to be a separate type of fishing gear from handlining in the International Standard Statistical Classification of Fishing Gear.

Target Species

Typical target species with handlining are demersal fishes like cod and snapper as well as squid. Trolling is mainly directed towards pelagic species like mackerel, tuna and salmon.

HAND LINES

As the name implies, it is operated by hand of the fishermen themselves. Although it looks simple in structure, the handline has many designs and modes of construction to effectively catch the target species. In general, a handline consists of a primary mainline, secondary mainline, hook and sinker. Other accessories include the wooden/bamboo/plastic spool for coiling the line, the swivel to prevent the line from twisting, and the stainless wire used to protect the hook from loss due to fish bites.

A simple handline has only one hook but to increase the chance of catch, several hooks are now common, especially for catching schools of sardines, mackerel, scad, sea bream, grouper and snappers. Hand lines can be used either for pelagic, mid-water or demersal species. Natural baits such as fish, squid and artificial baits such as plastic/silk materials which resemble a shrimp, squid or octopus are now used. In other areas, individual hook and lines are provided with plastic or bamboo floats and allow to drift with the current. The fishermen follow them for several hours and haul the floating hook and line with a catch.

The different behavior of target species has resulted in many modifications. In the tuna hand line, an iron rod measuring 50 to 90 cm. with 3 mm. diameter is embedded in a lead sinker of 1.5 mm diameter with a length of 15 mm. is used to increase the sinking speed of the line. Live baits such as frigate tuna and squid are used to increase gear efficiency.

The material of the hand line is mostly nylon monofilament with the mainline of 1.5 to 2.0 mm. diameter and the secondary mainline from 0.5 to 1.5 mm. Sinkers are lead while the hooks are manufactured by Mustad either using the J-hooks or ciecle hooks.

Fishermen also use single and multiple hooks for demersal fish. Nylon monofilament of 1.2 to 2.00 mm. are the primary and secondary mainlines. In the case for multiple hooks, branchlines of 0.45 mm. to 1.0 mm. are connected to the swivels of the mainline to prevent line twisting. Hook size and shape vary depending on the target species. The swivels of different sizes and shapes are made of brass. One peculiar innovation in the bottom handline is the provision of a bait bag where small fish or grounded fish are placed. Upon reaching the desired depth, the line is suddenly jerked to release the bait. Bait attracts the fish school into one area thus increasing the efficiency of the fishermen.

In Palawan, where coral reefs are abundant, commercial hook and line boats carry six to eight small non-motorized bancas (3-4 m) and take them to the fishing ground. Selected coral reef areas are assigned to each banca. Handlining is done the whole day and before sundown, the small banca are again gathered together while catches are weighed and stored in the fish hold. Crushed ice is used in preserving the fish catch. The mother boat stays on the fishing ground until the fish hold is full, or until provisions such as food, fuel and water are becoming limited. Target species are groupers, snappers and other coral reef fish.

Stainless steel clips are also used as connectors between the primary and secondary mainlines to eliminate kinks during hauling and entanglement of the lines due to catch. Likewise, swivels are added in the mainline to prevent twisting of the lines. Triangular stainless wire is also placed along the primary mainline to provide attachment for other branchlines. This innovation increases the number of hooks and finally the catch per unit effort. The use of silk cloth with varied colors as artificial bait has proven very effective for catching schooling fish. These are tied at the eye of the hook to cover the shank and the barb.

In General Santos City, tuna handlinning is very common as it is the major source of tuna "sashimi" for export. Several thousand handline boats (19 m) go out to the Celebes Sea, look for payaw and catch big-sized yellowfin and bigeye tunas. One trip usually lasts for seven (7) to ten (10) days including four (4) travel days. A major innovation introduced by the local fishermen is the coiling together of the baited hook and a stone weighing 0.5 to 1 kg. using 2-meters secondary mainline. This is then dropped near payaws until it reaches the swimming layer of tuna and the line is suddenly pulled to release the stone.

At this depth, fishermen jerk their lines frequently to attract tuna bites. Squid and frigate tuna are the preferred baits. The best bait should have a length of 15.24 cm. to 17 cm. for fish and 10 cm. to 17 cm. for squids. At present, the latest bait type which show increasing catch efficiency is the squid ink which is placed in small plastic and tied near the barb tip of the hook. When jerking, the tip of the hook make a hole of the plastic bag thus releasing a cloud of ink and tuna are attracted by the smell and intensify their feeding activity. The swimming layer of tuna under payaw ranges from 90 m. to 200 m. depth and fishermen must adjust their lines to this depth.

In some forms of handline, shrimp-like attractors with several hooks embedded in the body are used to catch big-fin reef squid. This operation is limited as the distribution of the reef squid is not very wide. Fishermen use bamboo pots instead to catch this squid.

TROLL LINES

Trolling is primarily used for surface and subsurface fish. Fish behavior towards splashing or rippling water produced by an object has led to some improvement in the hook and line techniques. Troll lines vary from region to region but use both natural and artificial baits. The artificial baits are hard plastic resembling a squid or octopus, while a cheaper material is multicolored silk cloth. Natural baits are fresh fish being caught during lighting.

The Nylon monofilament twine is the most common material for the mainline. The need for the hook to remain at the surface has introduced a modification by placing a foam float 1.5 meters before the hook. Often, stainless wire is part of the secondary main line which is placed before the hook. Troll lines are not only single hook but also use multiple hooks to increase the catch when encountering fish schools.

It is now common among fishermen to catch 5 or 10 fish in one hauling. Double hooks are also commonly used with artificial squid or octopus as baits to prevent the catch from escaping.

Troll lining can also be undertaken along the beach. Fishermen improvise a sort of wooden-kite shaped fish that when towed, produces ripples or a splash. Branchlines are tied along the mainline at 1 to 2 meters interval. The wooden-shaped fish is automatically reversed by towing it in the opposite direction. Fish caught are cavalla, caranx, and other species that frequent the shallow coastal areas.

Airplane-like splashers are also used in some places. The "saba-saba" which is made of wood and resembling an aeroplane is towed by a motorized banca near payaws to catch Spanish mackerel and tuna. Artificial baits such as squid and octopus-shaped plastic are also used. All the hooks have barbs to hold the fish catch. Two units are towed by the banca and the indication of fish catch is shown by rubber strips lengthening or snapping. The provision of a mechanism that produces water movement or disturbance along the sea surface using "saba-saba", fish-shaped kite and artificial baits make trolling an efficient method to catch surface swimming fish.

LONGLINING

The catching principle and construction of longlines.

Catching Principle

Longlining is based on attracting fish by bait attached to the hook. While handlining and trolling generally exploit the visual sense of the fish to attract it to the hook by artificial lures, longlining exploits the chemical sense of the fish.

Odour released from the bait triggers the fish to swim towards and ingest the baited hook with a high probability of being caught.

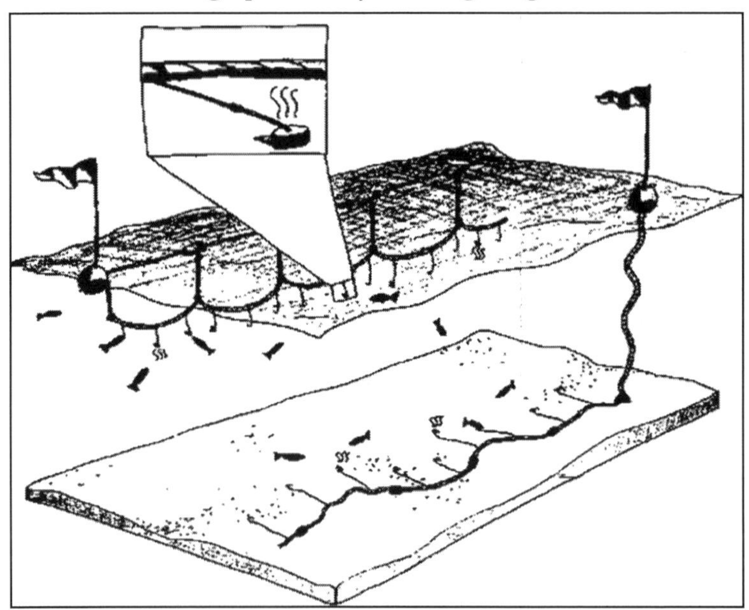

Fig. Catching principle and construction of longlines. Pelagic/drift (top) and bottom set (lower). Expanded view: baited hook connected with gangion (snood, branchline) to mainline.

Construction

As the name of the gear indicates, this is a long line (mainline) with baited hooks attached at intervals - connected to the mainline with relatively shorter and thinner leader lines (snoods, gangions). Depending on the type of fishery, there are great variations in the gear parameters, such as thickness and material of main and leader lines, the spacing between hooks, as well as hook and bait types.

Today, main and leader lines are almost exclusively made from synthetic materials like polyamide (nylon) or polyester. Multifilament (rope) is generally used for main and leader lines with demersal longlines (set on the bottom), while monofilament (gut) is commonly used in pelagic longlining. Hook type (size and shape) varies greatly with target species. Naturally, larger hooks and correspondingly stronger main and leader lines are used for larger fish. There is also a great variation in baits used in different longline fisheries, but the major types of bait are either different pelagic fish (e.g. herring, mackerel, sardine, saury) or different species of squid.

Operation

The longline fishing cycle includes the following main operations: baiting (threading a piece of bait on each hook), setting, fishing ("soaking" the line for some hours), retrieval, removal of fish and old bait, gear maintenance, baiting, etc. As with gillnets, the gear is basically the same in small and large-scale operations with the length of the line and number of hooks increasing with vessel size. Small, open vessels normally fish a few hundred hooks, while the largest longline vessels (LOA 50-60 m) may operate 50-60 km of longline and as many as 40-50 000 hooks per day.

With increased vessel size there is normally an increased degree of mechanised gear handling. Most longline vessels are equipped with power haulers. In so called auto-lining the laborious baiting process is also mechanised with machines that can bait up to four hooks per second as the line is set into the sea.

Target Species

Pelagic (drifting) longlines are typically used for catching species like tuna, swordfish and salmon, while bottom set longlines are used for demersal species like snapper, cod, haddock, halibut, ling, tusk, hake and toothfish.

LONGLINES

There are two major types of longline, namely; the bottom set longline and the tuna longline. The bottom set longline is the most common for small-scale fishermen as it is used to catch sparsely distributed demersal species. It is set in shallow muddy-sandy bottom to catch hairtail, grouper, snapper, shark and nemipterids.

The components of a longline consist mainly of floatline, mainline and branchlines. The branchlines consist of the mainline proper, swivel, wire leader and hook. The set longline is made of nylon monofilament. The twine and hook sizes depend on the target species. Usually, there are 200 to 1,000 hooks per set. Sliced fish is the common biat. Prior to setting, the hooks are baited and arranged in the hook rach. Sometimes, sand is placed in the box to prevent the lines from entangling. Operations maybe done early in the morning or before sundown. During setting, weights are tied along the mainline at intervals of 100 hooks or more to let the line lie on the sea bottom. A specific kind of catch is the spiny dogtooth shark which are available in some specific fishing grounds. Its oil has many uses. An improvised line hauler is utilized by the fishermen during hauling.

The original tuna longline is made of hard-laid nylon multifilament mainline with a combination of polyester rope hard laid and monofilament nylon in the branchlines.

Nowadays, the mainline and branchline are made of PA monofilament. The branchlines consists of the branchline proper, secondary branchline, swivel, wire leader and the hook. A flagpole with float, or a radio buoy are accessories used during fishing operations.

The tuna longline is operated by commercial fishing boats of 30 to 50 GT using Taiwanese-technology. The unit is divided into baskets or containers. The length reaches 20 to 30 miles long when set. Development in tuna longling is also very fast. Targeting bigger tuna especially the big-eyed tuna which inhabit the deeper water makes the fishermen increase the depth of the floatline and the branchline. Originally, the length of the floatline is 15 m. and the branchline is 20 m.

Nowadays, the floatling length is increased to 30 m. and the branchlines from 40 to 50 m. deep. Caught tunas are immediately stocked in the fish hold. In some companies, tunas are gutted, gilled and wased with sea water to remove blood and slime.

Live milkfish or frozen squid are used as baits. The shiny and moving fish bait attracts the tuna, hence, the catch rate has increased effectively. With still wide area of the Exclusive Economic Zones not exploited by commercial fishermen, tuna longline might be the method to increase production and exports.

GEAR NEEDED FOR TRAP AND POT FISHING

There are three things that you must have if you want to fish with traps and pots:

- Trap- and pot-making materials;
- Ropes and floats to mark, set and retrieve your gear;
- A vessel and equipment to operate the traps or pots and to transport the fish back to shore.

TRAP- AND POT-MAKING MATERIALS

The choice of trap-making materials depends on the type of trap you want to make and the fish you want to catch. You can construct many traps from locally available materials. For example, fish and lobster pots are often made from the sticks and branches of local trees, shrubs, mangroves, etc. If you are using mangrove wood to make pots, it is essential that you do not destroy the mangrove forests, as they provide important breeding grounds, shelter and food for many young and adult commercial fish and crustaceans. You can often use discarded material such as old tyres, plastic and earthenware piping to make effective traps.

Your choice of manufactured trap-making materials will depend on the local availability of these materials and on their prices. For example, if wire mesh is too expensive or unavailable, it can often be substituted by a piece of used nylon or cotton gillnet.

GOOD BAIT

Good bait is essential for effective trap fishing. Bait types vary according to the types of fish that you are targeting.

VESSEL AND ONBOARD EQUIPMENT

Vessel

The size of the vessel that you need depends on the type and number of traps you intend to use, the prevailing sea conditions in your area, the distance you need to travel to the fishing grounds, how long you want to stay at sea, the number of crew, etc.

Trap Handling Equipment

As pointed out above, many traps and pots can be operated by being pulled up manually, using the attached buoy rope. The use of trap/pot haulers and tippers on larger vessels using bigger traps can increase the efficiency of the operation.

Bait Storage

It is essential to find a suitable way to store your bait, both while you are fishing and onshore. Salted or dried bait only requires a dry place on the vessel. Fresh or frozen bait should be kept frozen or on ice or it will become rotten and unusable after a short time, depending on the temperature in which you are working.

Equipment for Handling and Preserving the Catch

To maintain the catch in good condition on board your vessel, you need to have the correct handling and preserving equipment.

This may include ice and ice boxes, freezers, brine tanks, filleting boards or tables, gutting knives and catch washing equipment (buckets, deck hoses, etc.).

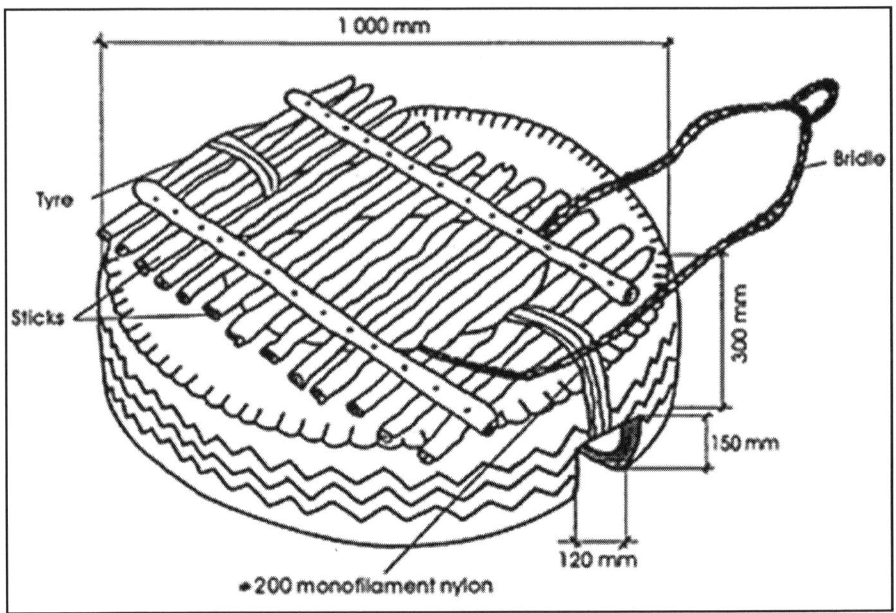

WHAT YOU NEED TO KNOW TO GO TRAPPING

Before you start trapping there are several things you will need to know:

- The type(s) of fish you want to catch and the type(s) of traps or pots that will catch them;
- The type(s) of bait needed for these fish and where you can get it;
- Suitable landing and storage for your catch on board;
- A market for your catch.

TYPES OF FISH TO CATCH AND THE TRAPS THAT WILL CATCH THEM

Although this manual gives general information on traps and pots, there are several other sources of local information on the fish that you wish to catch and the gear needed to catch them.

These include:

- The results from various types of fishing operations in the area;
- Private local knowledge from fishers, fishing companies and fresh fish markets and from fishing companies in other areas and countries;
- Public surveys by local fishing companies, the government, FAO or other international organizations.

TYPES OF BAIT AND/OR OTHER ATTRACTANTS NEEDED AND WHERE TO FIND THEM

Good bait is normally essential in successful trap fishing, although some subsistence fishers use only white or shining objects to attract the fish.

Sometimes the trap itself will lure the fish inside (e.g. habitat traps).

Bait must be:

- Effective in attracting fish;
- Readily available;
- Easy to store and conserve;
- Cheap enough to allow the operation to be profitable.

The most effective baits are usually oily fish such as sardines and mullet. Soft fish that break up after the trap has been set form a good "plume" that will readily attract fish. If traps or pots are set for longer periods, however, harder baits may be needed, or a combination of both soft and hard.

SUITABLE LANDING PLACES

When you start trapping, you will often land the catch on existing wharves or on the beach. Here you can sell it, pass it on to your family, ice it or preserve it by freezing, salting, drying, smoking, etc. If your trapping venture becomes well established and profitable, you will need to be sure that the catch can be landed easily, cheaply and in good condition. Although existing landing places may be good enough during the initial phase of your operation, you may need new or upgraded landing places as your fishery develops.

A MARKET FOR YOUR CATCH

If you wish to sell all or part of your catch, it is very important that you preserve it in a way that suits your customers.

The best method of preservation will vary according to:

- The time and distance from catching to the point of sale;
- The climate in the area where you fish;
- The value of the fish.

You should always try to land your fish in the best possible condition but you should also always balance the rewards of landing high-quality fish against the cost of achieving that quality. There is no point landing fish in first-class condition if it costs more to do so than the selling price you can expect to earn.

FUNDING FOR NEW VESSELS AND EQUIPMENT

The source of finance for fishing operations is outside the scope of this manual and is an aspect that can only be looked at locally. However, in some areas specific schemes are sponsored and funded by local organizations and banks. You should talk to your local fisheries organization about these schemes.

HOW TO MAKE VARIOUS TYPES OF TRAPS AND POTS

Many types of fish and other aquatic animals have been taken with simple traps and pots for hundreds of years. Of the main types of trap, this manual will concentrate on enclosed traps and pots.

HIDING PLACES OR HABITAT TRAPS

You can create artificial shelters and hiding places to attract and concentrate fish by placing bundles of branches, sticks or brush in the water. This method was probably used by ancient peoples and is still used in some areas of Europe, Africa and Asia. In more recent times, artificial reefs made from old car bodies, old motor tyres, concrete blocks, sunken vessels, etc. have been constructed to provide hiding places that concentrate fish. Floating devices for fish aggregating are sometimes used to concentrate commercial pelagic fish.

Brush Traps

You can make these traps simply by tying branches, sticks, brush etc. into bundles and either placing them flat or upright on the bottom or suspending them in the water. Fish and crustaceans sheltering in the brush trap can be caught by quickly lifting the bundles from the water. The bundles can be put in the centre of a woven basket to catch fish that try to escape when the trap is lifted. Alternatively, you can put a dip net under the bundles while you are lifting them. Brush traps can be set in long rows or on longlines for easy hauling.

Eel Tubes

Eels and other long thin fish can be caught by setting tubes for them to use as shelter. Such tubes are often closed at one end and you can make them from bamboo, hollow logs, steel, plastic or earthenware piping, etc. They are usually tied into groups of two. Bait can be placed inside the tubes to attract the eels or fish. When lifting the tubes, you must take care that the eels do not escape. In some places divers seal the ends before the tubes are lifted.

Octopus Pots and Traps

Traditionally, unglazed earthenware pots were used in the Mediterranean and Asia to catch octopus. However, you can use other materials, including plastic, steel or earthenware piping, old motor tyres and empty large mollusc shells. Octopus will enter almost any receptacle that provides shelter.

BARRIERS TO FISH MOVEMENT

Simple barriers made from stone or wood were probably one of the first types of trap used by prehistoric peoples. Many such barrier traps (fish screens, labyrinth traps, corrals) are still used today throughout the world. Each area has its own particular designs for permanent barrier traps, based on the types of fish in that area, their behaviour and migration and the area being fished.

These traps are normally constructed with a long arm or leader, usually from the shore outwards, with an arrowhead-shaped trap on the outer end. The design of this trap varies widely in its complexity according to the species of fish being targeted and their movements.

ENCLOSED TRAPS AND POTS (BASKETS)

Enclosed transportable traps and pots (which are often referred to as "pots") are by far the commonest types of trap used throughout the world. Although there are many types and shapes, including rectangular, circular, hexagonal, conical, semi-cylindrical, chevron-shaped and heart-shaped (arrowhead), the way they work is the same: fish enter the trap by one or several entrances or funnels and are prevented from escaping.

These are:

- Frame;
- Outer covering;
- Entrance funnel;
- Bait holder;
- Door to empty the pot or trap;
- Escape gap;
- Ballast;
- (Sometimes) a corrosion anode when the frame is metallic.

Frame

Frames are made from strong materials that prevent the traps and pots from losing their shape during fishing and storing. Trap frames are often made from steel rods, although sawn timber and strong sticks are used in some places. In northeastern Brazil, rock lobster pots are traditionally made from mangrove and other sticks. In Canada and northeastern areas of the United States, while lobster pots were traditionally made from sawn timber, now more plastic-coated wire is used. Beehive pots are usually made from flexible material such as cane or bamboo, often with a base of bent steel rods. In Australia and New Zealand, pots and traps are now often made entirely from welded steel mesh that does not need a frame to support it.

In some cases, the frames are made so that pots can be folded in order to save space when they are stored on the deck of the boat.

Covering

Modern rectangular traps and pots are covered with wire netting, nylon mesh, plastic-covered steel, welded steel mesh, etc. The choice of material depends on traditional usage, availability and cost. Tests to see if there are any differences in catch rates between wire and nylon mesh have shown that in most cases there are not.

In some areas, bamboo or plaited cane is still used. Beehive pots are commonly made from cane or flexible branches without a frame, although in some areas they are made with a solid frame and covered with cane, wire or nylon netting.

Moulded plastic pots appear to be cheap and easy to stack and there is no reason to believe that their catch rates would be inferior. If you are fishing in an area where sharks are common you should use steel mesh rather than nylon mesh.

Funnels

Although beehive pots have only a single funnel at the top, other trap and pot types may have several funnels. The inner ends of funnels are often directed downwards ("horse-neck") or constricted in some way to prevent escape. Some pots, mainly for lobsters, also have the "bedroom and parlour" design. This consists of a funnel or funnels leading from the outside to an initial chamber (the "parlour") and a second funnel to an inner or holding chamber (the "bedroom").

Research has shown that traps with more than one funnel catch more than traps with a single funnel, although the quality of the catch is often reduced as a result of injury to the fish caused by the increased number of projections on the inside of the trap.

Door

A door is usually placed in the main body of the trap to make it easy to remove the catch. Most traps have a properly made door. A few have an area where the outer mesh can be opened and closed at an edge near the frame. Lobsters and rock lobsters caught in beehive pots are normally taken out through the funnel at the top.

Bait Holder

Bait is normally secured in the catching chamber of the trap or pot. Large, solid pieces of fish, animal bone, small land animals, etc. are tied into the pot with wire or twine.

If small pieces of bait such as small fish (e.g. sardines or sprats) or chicken heads are used as bait, they should be placed in a bait container made from wire, plastic or synthetic netting to hold them together and in place. In some cases, the bait is put in a plastic or metal container with a few small holes punched in it to save the bait. This can only be done if the bait is extremely attractive to the fish you want to catch.

In some fisheries traps are not baited, as fish or squid use the pots as shelters. This is especially the case in Caribbean trap fisheries, where the reef fish that are targeted use the traps for shelter.

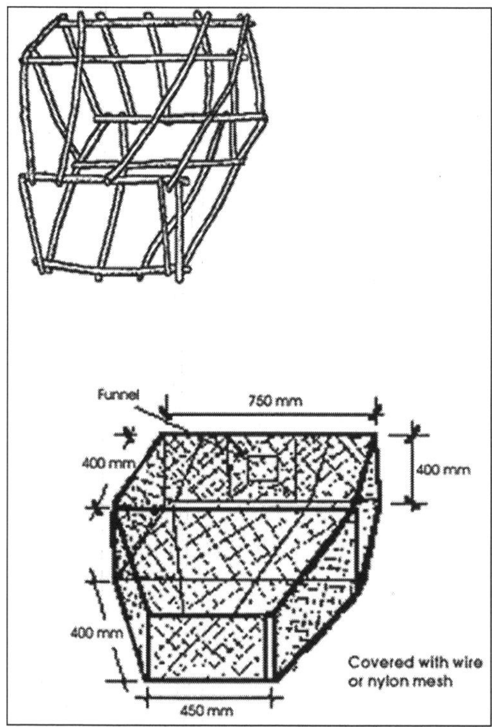

Fig. Rock lobster pot made from mangrove sticks (northeastern Brazil)

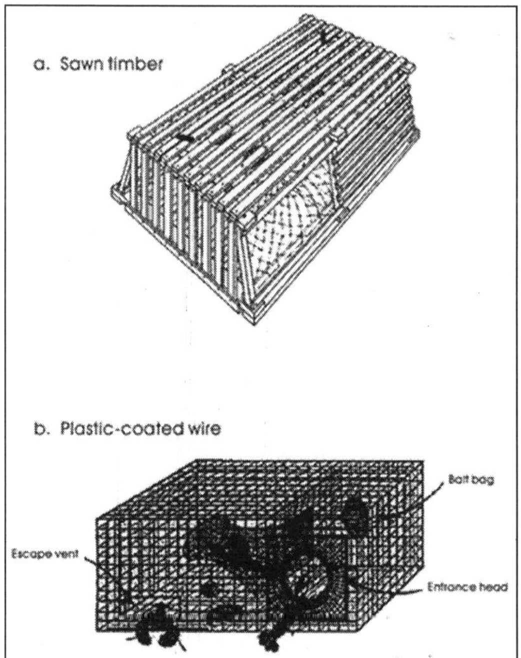

Fig. Lobster pots used in northeastern United States.

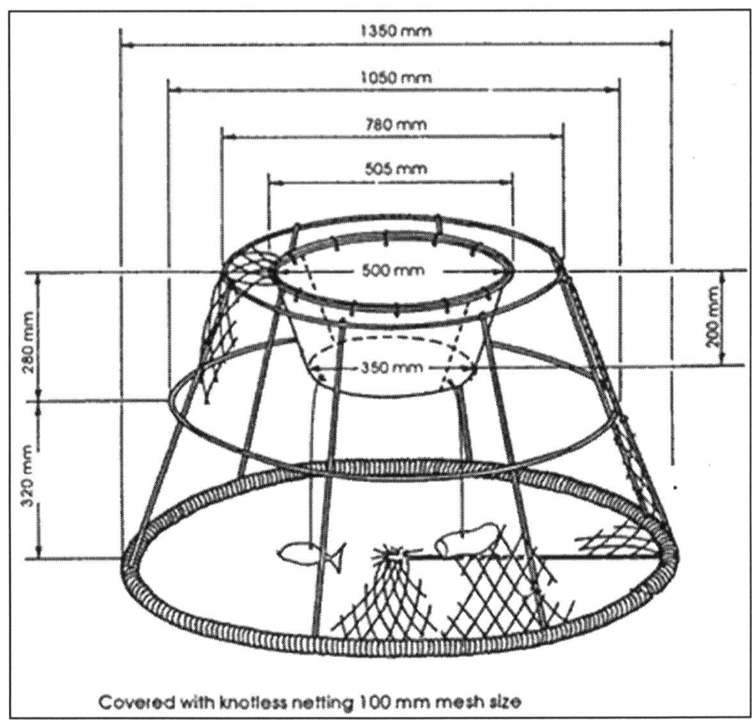

Fig. Beehive pot type, cone iron frame, 8 mm diameter (used in Japan for deep crab fishing).

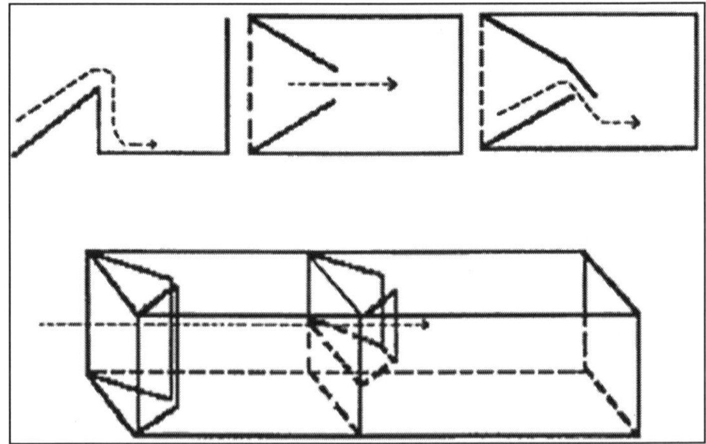

Fig. Diagram of different funnel types.

Escape Gaps

Escape gaps are often fitted into pots and traps to make sure that undersize fish or crustaceans, especially lobster and rock lobster, are not taken. Government regulations have often been passed to ensure that escape gaps are used to let small fish and crustaceans escape. In many rock lobster fisheries, including Australia, New Zealand and Cuba, escape gaps are compulsory in all pots to allow the escape of juveniles. In Australia and New Zealand, rock lobsters with head (carapace) lengths less than 76 mm must be released or allowed to escape.

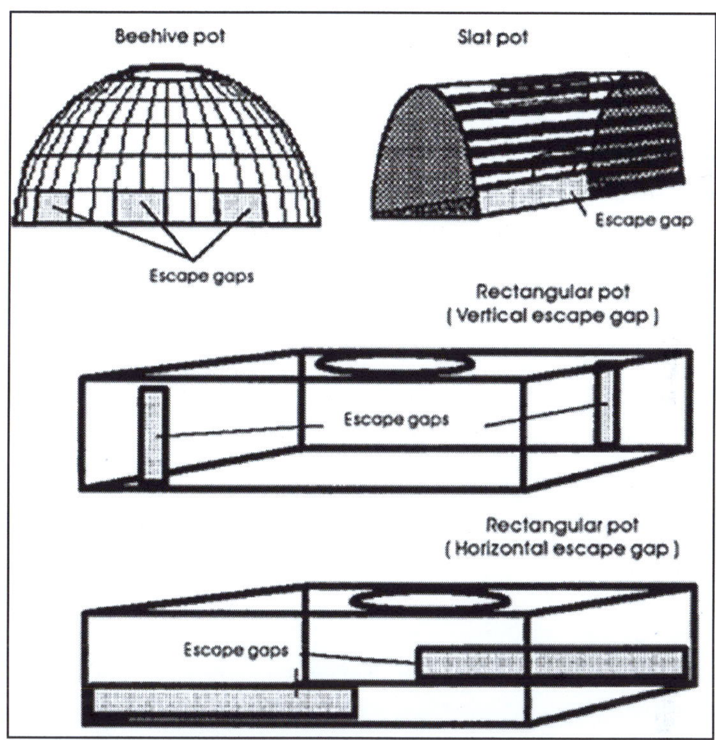

Fig. Escape gaps for lobster pots.

Ballast

Weights or ballast are often placed in pots and traps before setting to prevent tidal flows and currents from moving them from where they are set. This is especially the case for traps and pots made of wood or other light material. Weights may be concrete blocks, steel bars or other heavy material such as bricks, stones or rocks. Strategically placed ballast may also help the trap to land the right way up.

Anodes for Metallic Framed Pots or Traps

Because of the corrosive effect of seawater on steel and other metals, an anti-corrosion anode is often attached to steel traps and pots to extend their useful life. Anodes are usually made from a block of zinc with a wire through it to tie it to the pot or trap. They are not cheap and should only be attached to metal traps and pots that are costly to make. Galvanized wire netting also reduces the rate of corrosion in seawater. If the ongoing cost of replacing traps and pots that have been damaged by corrosion is high, you should consider using plastic mesh for the outer covering of your traps and pots and plastic-covered metal for the frames.

MAKING TRAPS TO CATCH DIFFERENT TYPES OF FISH, CRUSTACEANS AND CEPHALOPODS

As different species react in different ways to traps and pots, we will look at how traps can be made for three important target groups.

The groups we are considering are:

- Finfish;
- Lobsters, shrimps, prawns, crabs and other crustaceans;
- Octopus, squid, cuttlefish, etc.

Finfish Traps and Pots

Pots and traps for taking finfish are used in most parts of the world and there are many types and variations. To simplify this manual we will look only at some common types and their more popular or innovative variations.

Traps and pots are frequently used to take fish in areas where coral or the rocky bottom does not allow the use of other types of fishing gear. For example, in the Caribbean traps are the primary fishing gear. In many other areas a large number of fishers use simple traps made from mangrove poles covered with wire mesh.

The most common types of finfish traps used throughout the world include:

- Caribbean traps (arrowhead, “Z”, “S”, etc.);
- Round traps;
- Rectangular traps;
- “D”-shaped traps;
- Collapsible traps;
- Pelagic fish traps;
- North Atlantic cod pots;
- Plastic multipurpose traps.

Caribbean traps (arrowhead, “Z” and “S” traps).

The four most common fish trap designs used in the Caribbean are the:

- Antillean “Z” trap;
- “S” trap;
- Arrowhead or chevron trap;
- Rectangular trap.

Arrowhead (or chevron), “Z” and “S” traps are also used in Asia and the Mediterranean. They are principally used in tropical waters and have been successfully used to take emperors (Lethrinidae), snappers (Lutjanidae), cod (Epinephalidae), bream (Sparidae), parrot fish (Scaridae), goat fish (Mullidae) and trevallies (Carangidae).

The most common of these designs is the Antillean "Z" trap. This takes the form of a double chevron or "Z" with two down-curving "horse-neck" entrance funnels. Typically, these measure between 180 and 230 cm in length and are 60 cm high. In the Caribbean, specialized artisans construct these traps in a number of stages.

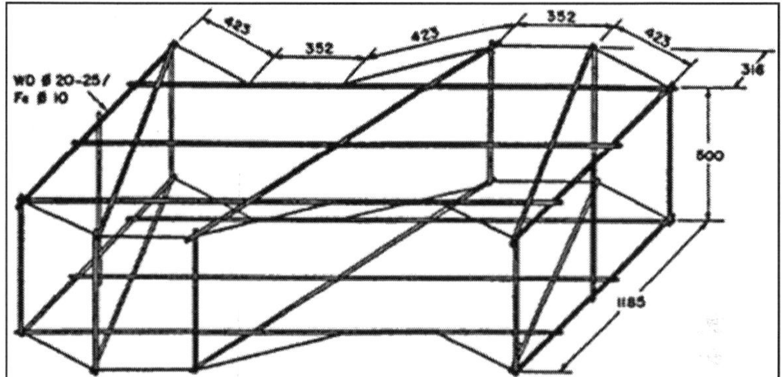

Fig. Typical Caribbean "Z" pot: frame and dimensions.

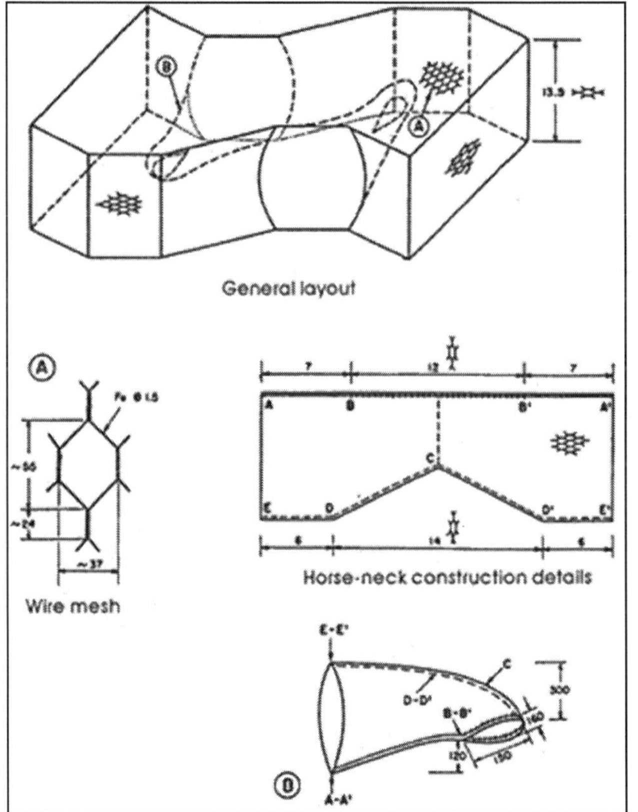

Fig. Typical Caribbean "Z" pot: details for construction.

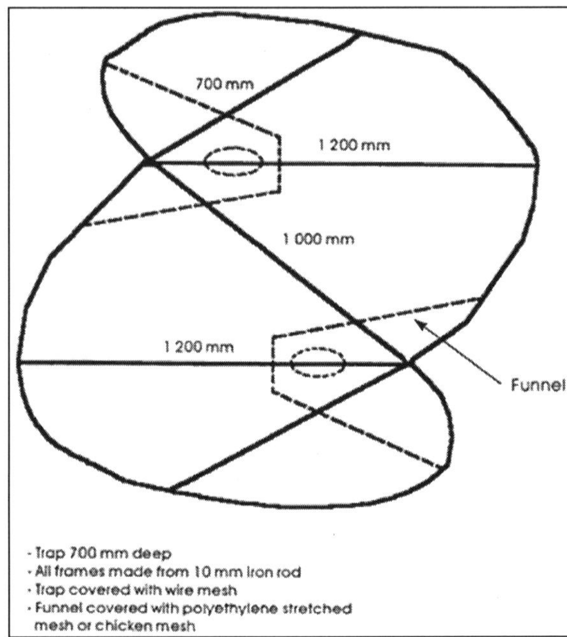

Fig. A typical "S"-shaped trap.

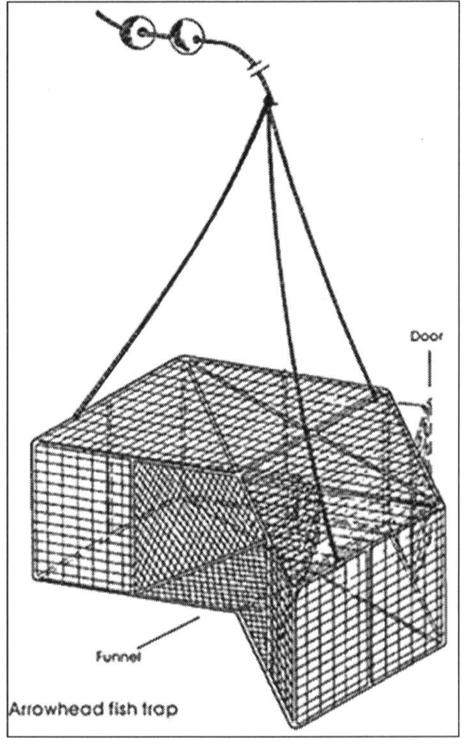

Fig. Arrowhead or chevron fish trap.

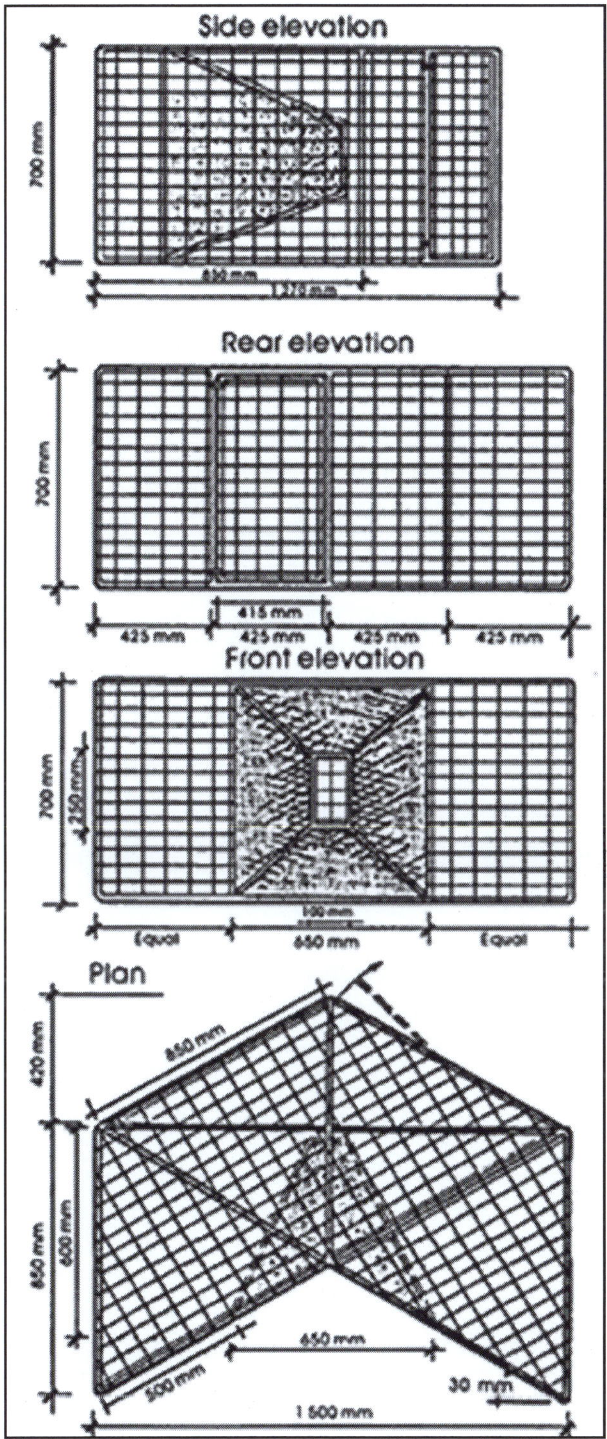

Fig. ″ Design details of the arrowhead fish trap.

Making finfish traps. If you want to construct a "Z" trap, the materials that you need are:

- A 50 × 1.2 m roll of galvanized hexagonal-weave wire mesh;
- 24 hardwood sticks or poles about 2.0 m long;
- 1 kg of 3 cm nails;
- *Tools:* Hammer, saw, wire-cutting pliers and machete or hatchet.

With these materials you can make four pots 270 cm long, five pots 240 cm long or six pots 210 cm long.

The essential stages in making a typical 240 x 120 cm Antillean trap are:

1. Cut the roll of wire into 240 cm lengths.
2. Cut the overall "Z" shape into the wire. Each "Z"-shaped pattern will be the top or bottom of a pot.
3. Cut two lengths that are the height of the pot to be constructed; 120 cm in this case. These are the sides of the pot.
4. Set the sides on the ground and attach the top and bottom, forming a pot without a funnel or wooden supports.
5. Cut the two funnels, using 150 cm lengths of wire for each.
6. Double over the funnel material to make the two "horse-neck" or down-curving funnels; burr the edges.
7. Fit the funnels into the opening cut in each corner of the pot and form the downward curve of the "horse-neck" by hand.
8. Fit a 30 cm "V"-shaped or square door in a corner of the pot.
9. Fit the pot sticks and wire them tightly into place to make up the framework in the previously formed pot. Leave a 10 cm overhang at the lower ends of the vertical sticks to hold the pot off the bottom.
10 Fit three long "key" sticks to the whole length of the top and bottom of the pot. The "top-back" and "bottom-back" sticks are important, as they are used as attachments for the buoy ropes.

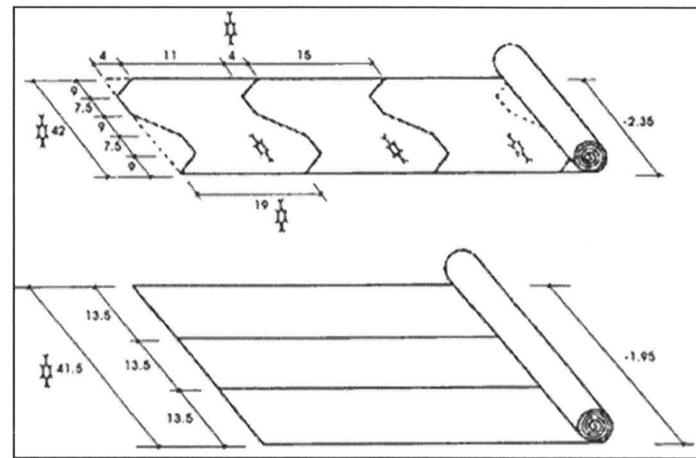

Fig. Diagram for cutting rolls of mesh for a typical Caribbean "Z" pot.

The mesh used in Caribbean fish pots is usually galvanized hexagonal-weave chicken wire with 4 cm ($11/_4$ inch) openings. In some parts of the Caribbean, mainly Jamaica, finely woven bamboo cane is used to replace the wire over the wooden frame. Steel reinforcing bars are sometimes used to make pot frames.

Changes to the trap design and modification to the entrance funnel have been found to make big differences to the catch rates of these traps. Straight funnels seem to have higher catch rates (and escape rates) than those with a "horse-neck". "S" traps were found to outfish "Z" traps by a factor of about 25 percent, while both were found to perform better than arrowhead traps.

Recently, modified arrowhead traps have been introduced into Australia. The trap frame is made from 10 mm mild steel which is covered with 50 x 75 mm rectangular 1 to 3 mm diameter mesh. An outward-opening door is placed in the rear of the trap. The details of the design of these traps. The buoy lines are made from 12 mm rope adjusted to one-and-a-half times the depth of the water being fished. The traps are set with the point towards the tide.

Round traps. As with other types of trap, the size and height of round traps vary with the area being fished, the type of fish being targeted and the availability of trap-making materials. The number of funnels in round traps varies, but one, two or three are usual. Additional funnels can increase the catch rates by allowing fish to enter the trap more quickly, but there is a higher escape rate and the additional internal projections from the extra funnels cause greater damage to the fish in the trap.

The western Australian snapper trap is a typical commercial round trap. It is also popular in northern Australia and has been used in the development of the tropical snapper fishery in that area. The dimensions and construction details of these traps can vary. A version of the western Australian trap has three funnels and a frame made from 10 mm mild steel rod. This frame is 180 cm in diameter, 70 cm high and has three braces top and bottom to support the funnels and the attached buoy ropes.

In this trap the funnels are straight, tapered to a 25 × 11 cm opening. The frame is usually covered with welded wire mesh of about 75 × 50 mm but it can be covered with other types of wire, heavy nylon netting, etc. Similarly, the funnels can be made from steel rod and covered with heavy nylon mesh or the same material as the body of the trap. In some traps, the funnel is shaped from welded mesh without a frame. A door is normally placed in the side of the trap to make it easier to take the catch. As with other traps with more than one funnel, the catch rates increase with the number of funnels but damage to the fish is increased by the extra projections from the funnels inside the trap.

The main fish caught in tropical Australian waters with round traps are tropical snappers (Lutjanidae), emperors (Lethrinidae) and groupers

(Epinephlidae). A simple round trap can be made from a horizontal cylinder of wire netting with a wire mesh funnel at one end or funnels at both ends. A similar trap can be made of nylon mesh covering a steel frame constructed from hoops and steel rods. These simple traps are often referred to as "drum nets". They are used in rivers, streams and lakes to catch fish moving close to the bank. A freshwater fish pot or drum net used in Germany for catching tench, roach and carp.

Drum nets are normally set by rolling them into the water from the bank and down the underwater slope to the preferred fishing place. A line attached to the trap winds round its centre as it is rolled down the slope. The trap is taken from the water by pulling on the line, which rolls it back up to the bank. Drum nets can either be baited or placed unbaited in an area where fish swim in currents.

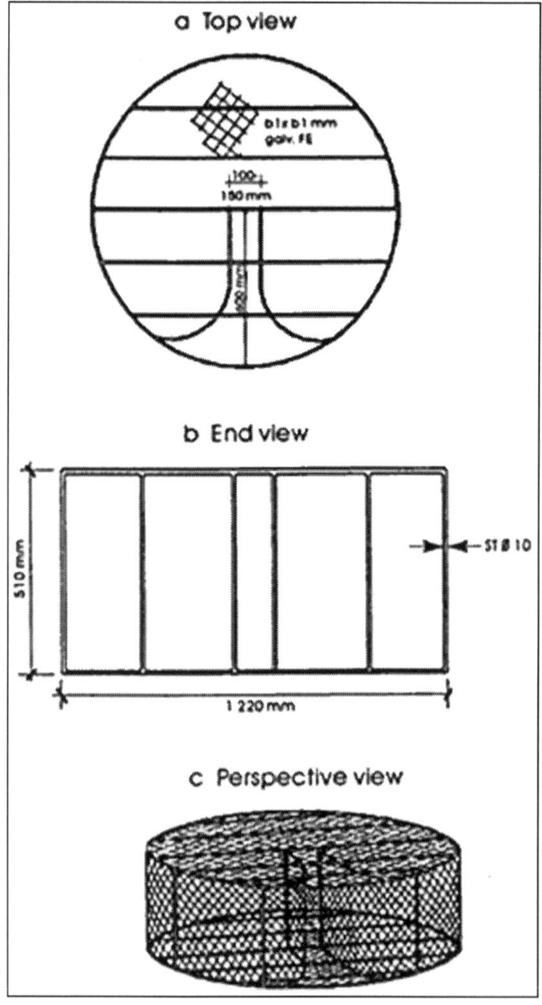

Fig. Round fish trap for snapper (Western Australia).

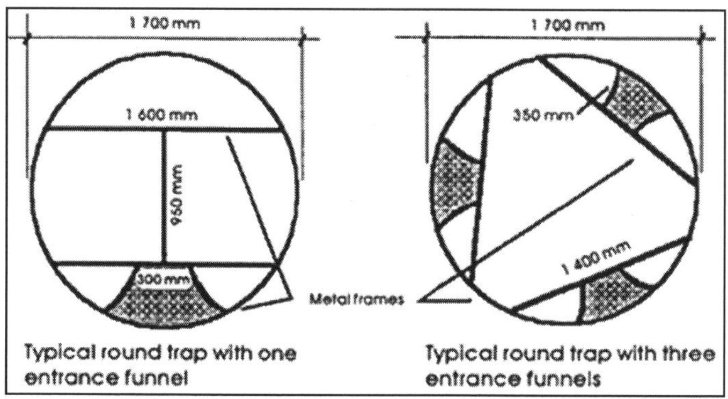

Fig. Positioning of funnels in round trap.

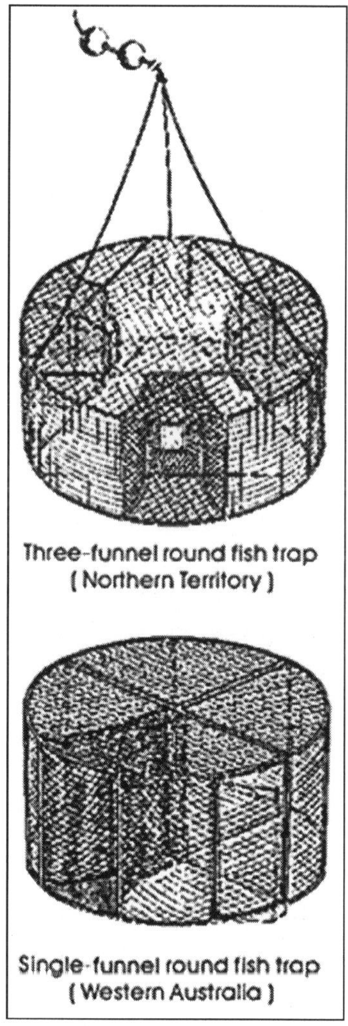

Fig. Australian round traps.

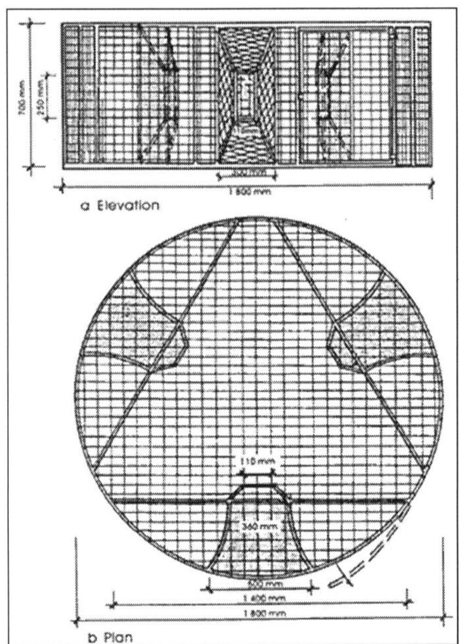

Fig. Three-funnel round trap from Australia.

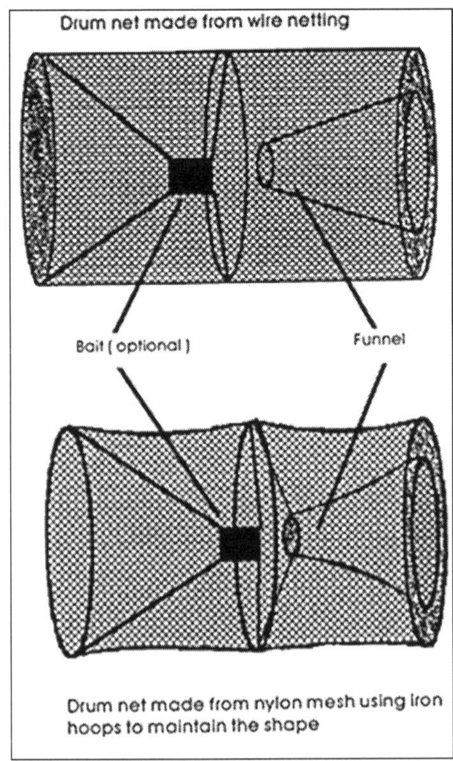

Fig. Drum net type of trap.

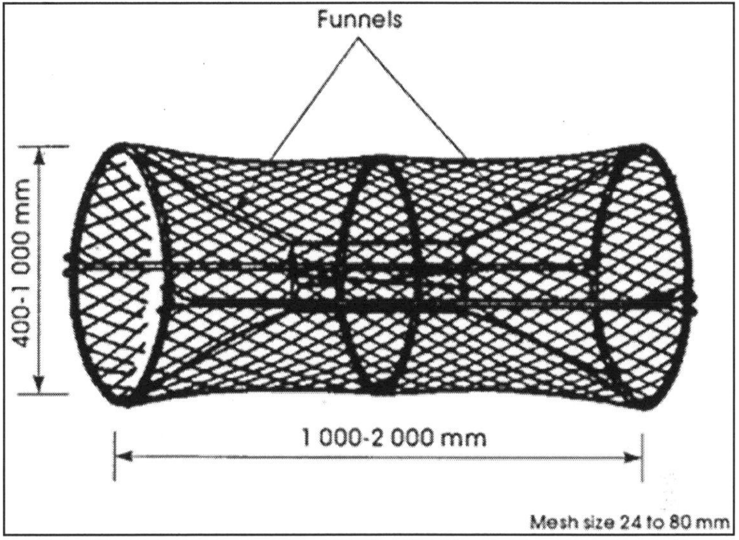

Fig. Drum net fish trap for inland fisheries (Germany).

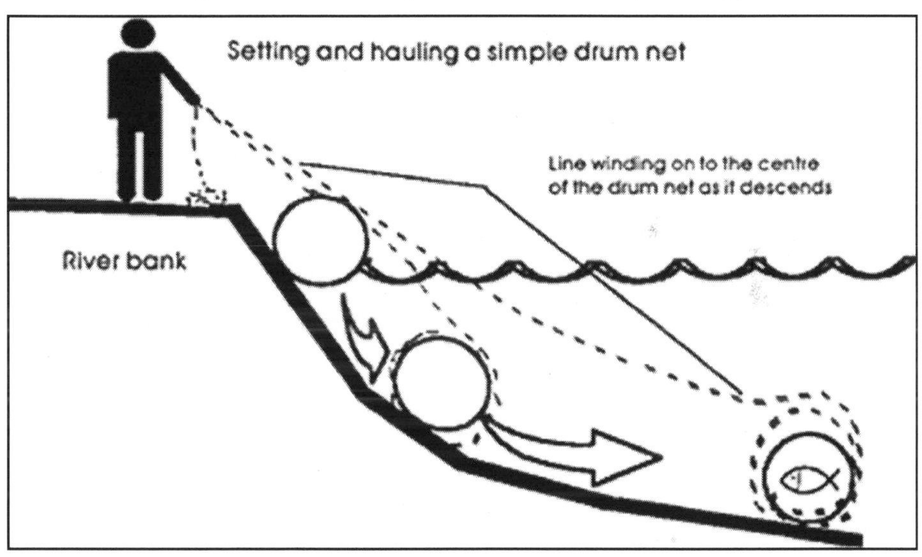

Fig. Setting and hauling a simple drum net.

Rectangular finfish traps. Rectangular traps vary greatly in size and construction material, but modern traps are usually made from steel rod or welded mesh. Because of their tendency to roll, they are used in areas where the current flow is not great. Rectangular trap frames of different sizes are made from dressed, rough or natural timber, including mangrove branches. Wooden traps are used to take snappers, sea bass and many other bottom-dwelling species in temperate and tropical waters.

Rectangular traps are very common throughout Australia and have been modified to fish in various conditions. They are typically constructed from mild steel and are covered with welded mesh, chicken wire, wire netting or nylon netting. A typical wooden trap is about 2.0 × 1.5 × 1.0 m overall and constructed from 50 × 25 mm timber bolted together. Chicken wire is then stretched tightly over the frame, giving the trap further strength. It has a funnel placed at one or both ends, also made from chicken wire or from a frame covered with nylon netting. Heavy ballast is positioned in the bottom of the trap to sink it and to hold it on the bottom.

A rectangular trap often used in southern Australia is made from a single sheet of 50 × 50 mm welded mesh cut and bent into shape. The bait is tied or placed in a bait basket (or bait saver) in the centre of the trap. A baffle is often placed in front of the funnel to prevent fish from escaping back through it. A modified rectangular trap made from a sheet of 50 × 50 mm welded steel mesh has been developed in Australia. This trap is 32 × 32 meshes (i.e. 1.6 × 1.6 mm) square, with rounded corners. The sides and funnel are bent from a piece 15 meshes deep. The mesh is supported by a frame made from 13 mm round steel bar.

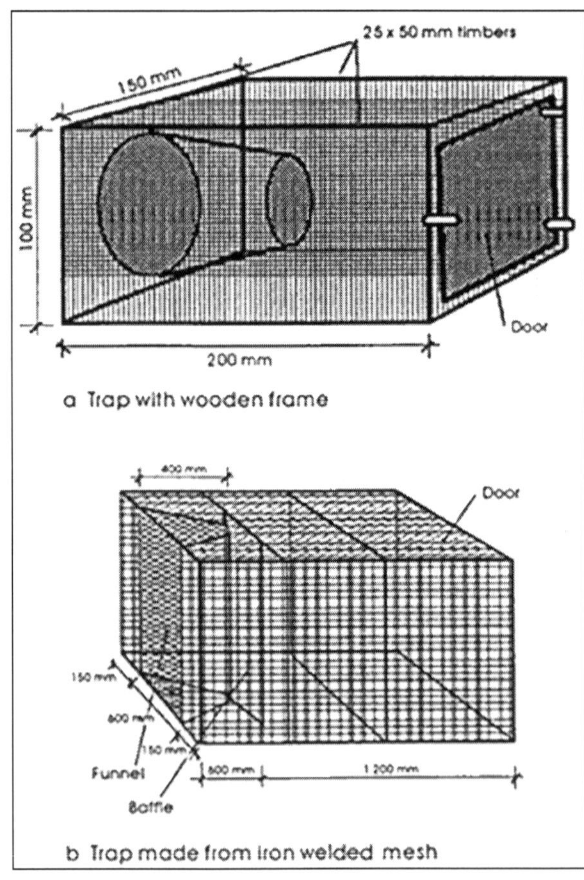

Fig. Rectangular fish traps.

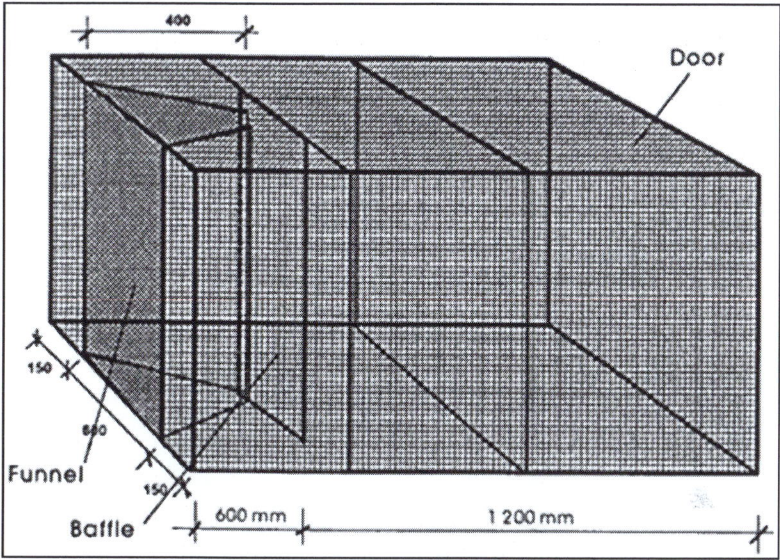

Fig. Welded mesh fish trap (South Australia).

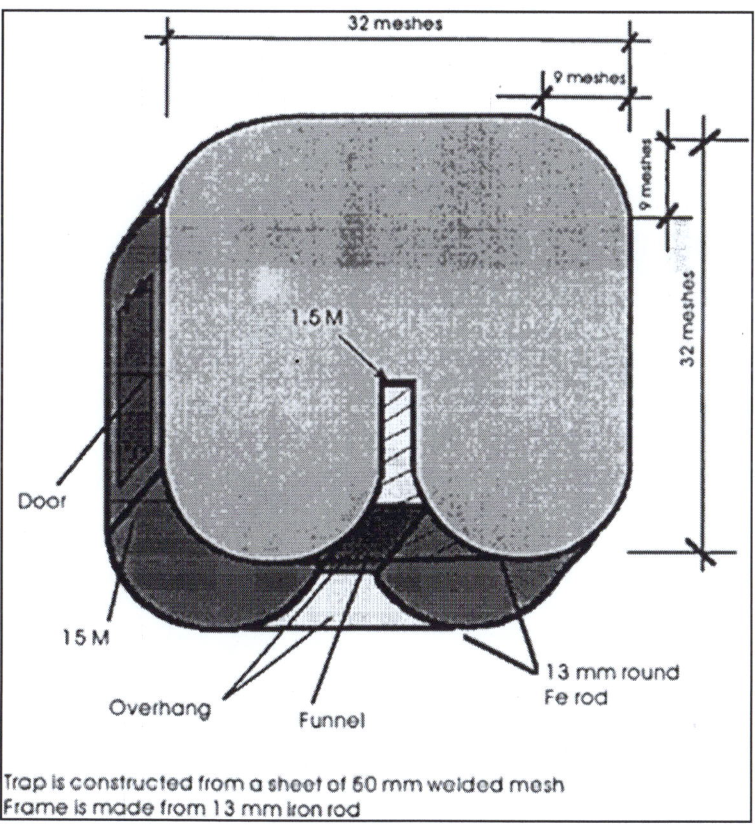

Fig. Australian welded mesh fish trap.

"D"-shaped finfish traps. The "D" shape is preferable to the traditional rectangular fish trap in areas of stronger currents, as it offers less resistance to water flow and is less likely to roll.

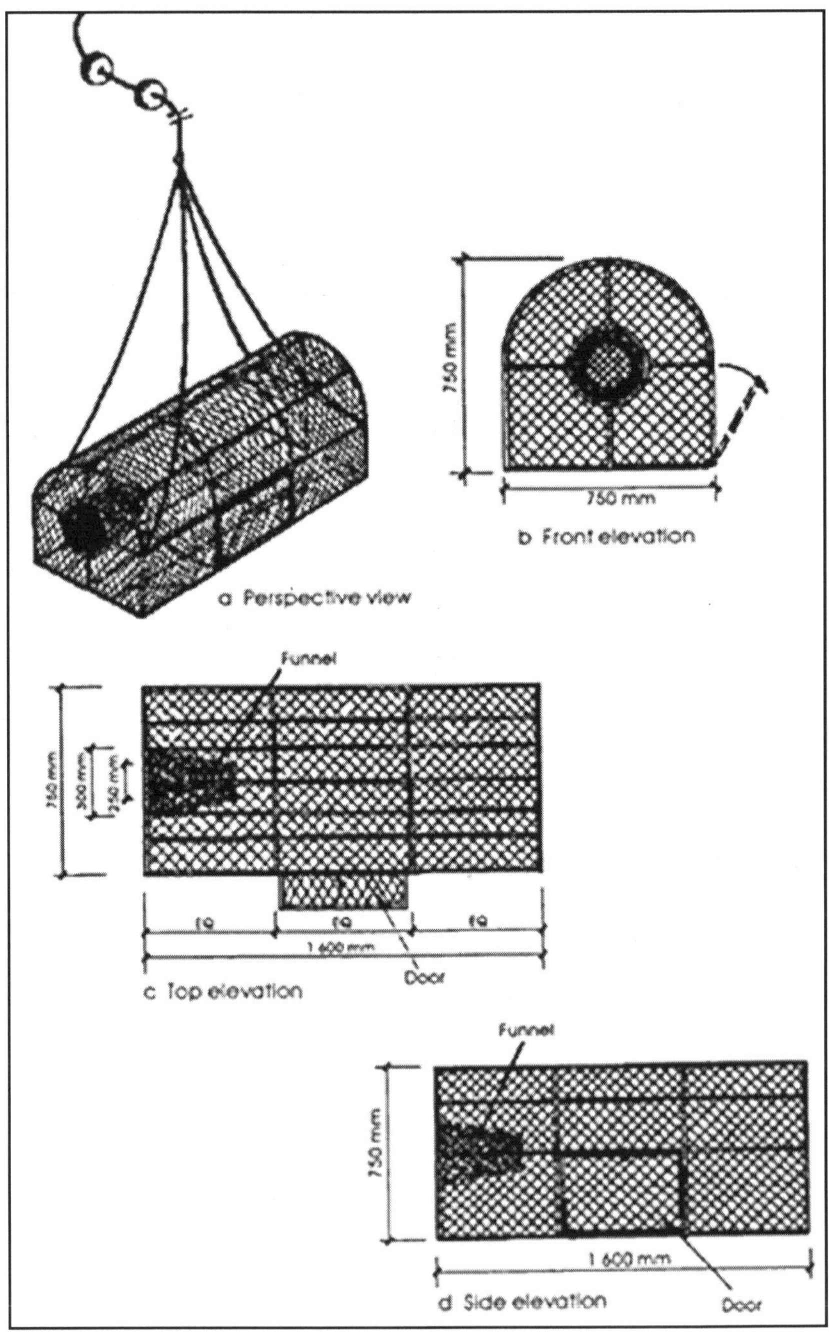

Fig. "D"-shaped fish trap (Australia).

Although "D"-shaped finfish traps are common throughout the world, the design illustrated in this manual is one that was modified for Australian conditions. This trap is 160 cm long and 75 cm high. It has a funnel at one end and a hinged door at the side for baiting the trap and removing the catch. Typically, the frame is welded together from 12 mm round mild steel, although other materials such as wood are occasionally used. The traps are usually covered with chicken wire, but welded mesh or prawn netting can also be used. The funnel tapers from 30 to 25 cm. It is made of similar material to the main trap and is fixed into the centre of one end. The internal funnel opening can be oval-shaped with the longest axis vertical.

These traps are used in southern Australia to take leatherjackets (Monocanthidae) in deeper waters of the continental shelf and pink snappers (Pagridae) and other bottom-dwelling fish in shallower waters. In tropical Australia they are used to take fish such as jacks and trevallies.

Collapsible finfish traps. The collapsible finfish traps are developed in the Northern Territory of Australia to fold up when not in use, thus increasing the number of traps that can be transported on board vessels to the fishing grounds. They are showing considerable promise in the developing fishery off northern Australia.

All frames are constructed from 6 to 10 mm diameter steel bars. Traps are covered with chicken wire, welded mesh or shrimp netting. The panels are constructed from similar steel bars and covered with galvanized wire mesh (50 mm diamond). This mesh is laced on to the frames with 1.20 mm tie wire. The round entrance funnel is also constructed from 50 x 50 mm wire mesh. The entrance is laced to the frame with twine or light tie wire, which may be more durable.

The mouth of the entrance is made from a 25 to 30 mm diameter disc cut from 100 mm PVC piping. This disc is softened in boiling water and formed into an oval shape to approximate the shape and size of the target fish. Two holes 25 mm apart are then drilled in the disc, which is cut between the holes. A ring is laced through the last meshes of the funnel and the cut closed with tie wire or twine. The funnel is held open by shock cords running from the PVC disc to the top and bottom panels, with enough slack to collapse the trap. A hauling yoke made from 14 to 20 mm rope is spliced on to the side of the top panel at the end of one of the cross members. The target species and operation of these traps are similar to those described for other finfish traps.

Pelagic fish traps. Although they are not yet widely used, these traps are effective in some areas for taking pelagic fish. They work on the same principle as a fish aggregating device, by attracting pelagic fish to the "protection" that they provide. The trap illustrated in this manual is used in southern Australia to take mid-water fish such as king fish, rainbow runners, trevallies and jacks (Carangidae). It is constructed in the same manner as the wooden trap but without the weights. In addition, a marine ply roof is fitted to the trap and painted black. The shade created by this roof lures the fish into the trap for shelter, so no bait is needed.

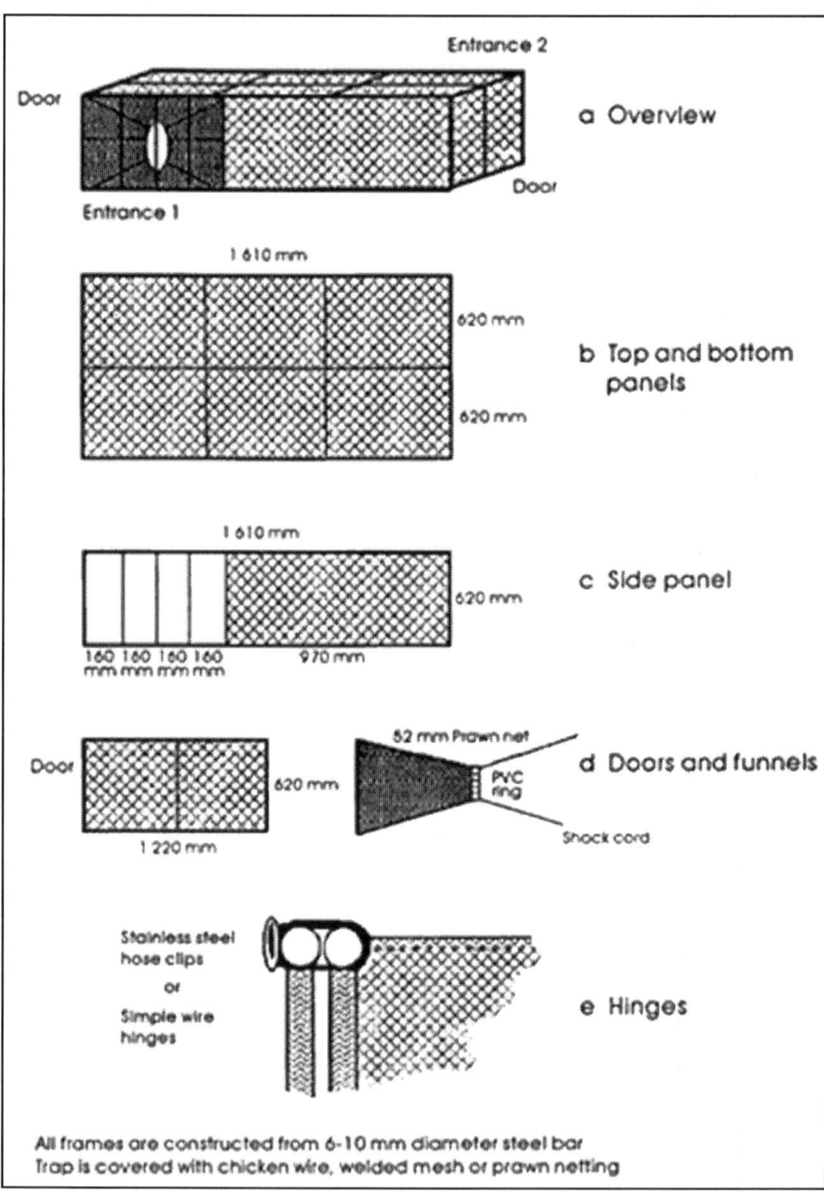

Fig. Collapsible finfish trap (Australia).

The method of setting these traps. When the trap is hauled up, it is tied to the side of the vessel and fish are taken from it with a scoop net before it is brought on deck. North Atlantic tusk and cod pots. Tusk or torsk (Brosme brosme) have been caught in pots off the Norwegian coast for a number of years. Norwegian technologists have developed an effective pot for taking live cod, following a recent trend for fishers to take these fish for the high quality of the end product. A requirement for an alternative to gillnets has helped the development. The pot has two chambers and is somewhat larger than the traditional tusk pot. It has two fairly wide entrance funnels leading into the lower chamber, with a narrow entrance leading to the upper chamber. A bait bag or saver is fixed in the lower chamber between the two funnels. The pots are baited with squid and set on a string or longline at depths varying from 50 to 300 m.

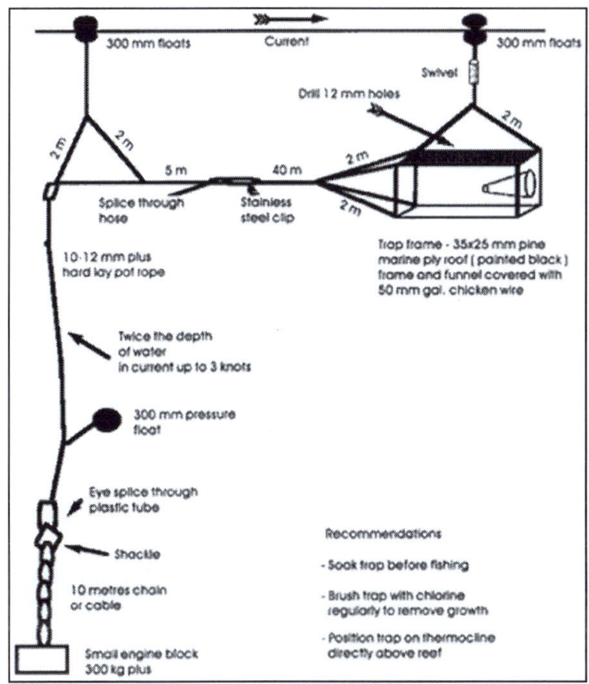

Fig. Pelagic fish trap.

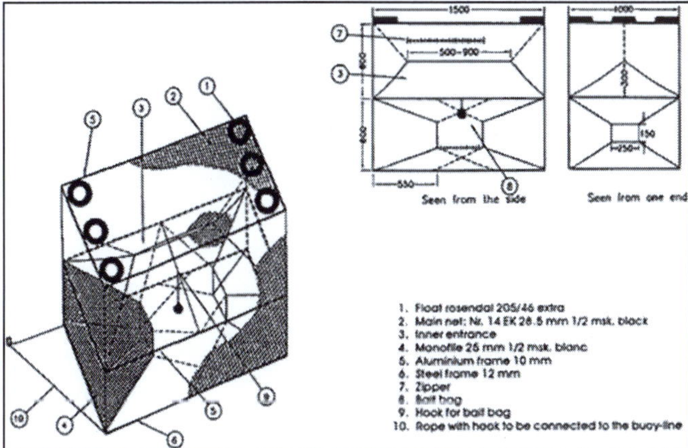

Fig. Norwegian two-chamber pot for cod.

Plastic multipurpose finfish/lobster traps. A number of trap and plastics manufacturers are experimenting with multipurpose traps and they have been tried out in many fisheries. They are not considered in this manual as they cannot be constructed from readily available local materials and require expensive plastic moulding equipment.

TRAPS AND POTS TO CATCH LOBSTERS, SHRIMPS, PRAWNS, CRABS AND OTHER CRUSTACEANS

Shrimp pots. The terms "shrimp" and "prawn" refer to different animals in different parts of the world: larger types being referred to as prawns and smaller types as shrimps. However, the common tropical shrimps (Penaeidae) are referred to as shrimps in most places. Temperate types (Pandalus and Palaemon) are usually called prawns. In this manual we will conform to this usage as far as possible. Several types of pot are used for prawns in the northern Atlantic. The details of a square wooden pot to take prawns (Palaemon) from the English Channel and a round pot to take prawns (Pandalus) on the western coast of France. The most successful commercial fishing method for tropical prawns and shrimps, both shallow- and deep-water, is trawling. Some preliminary experimental trapping of deep-water prawns (Hetrocapus spp.) has been carried out off Thailand, eastern Australia and some islands in the Pacific. Recent experiments in northern Australia used a number of trap designs for penaeid shrimp with little success.

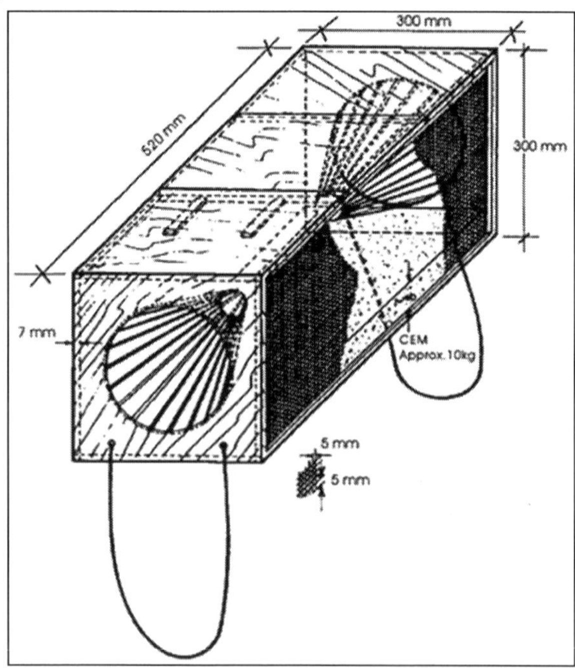

Fig. Prawn pot used in the English Channel.

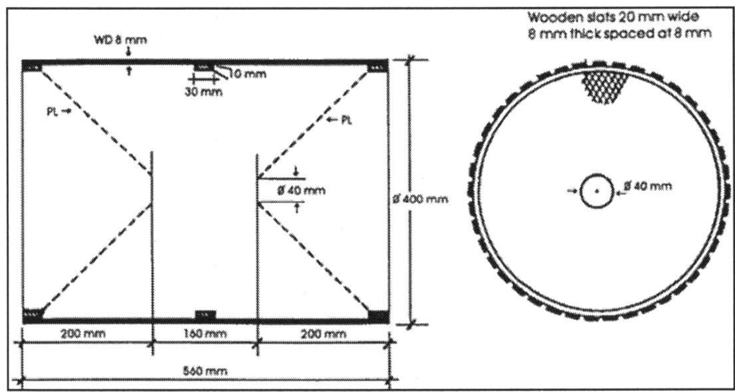

Fig. Shrimp pot used on the western coast of France.

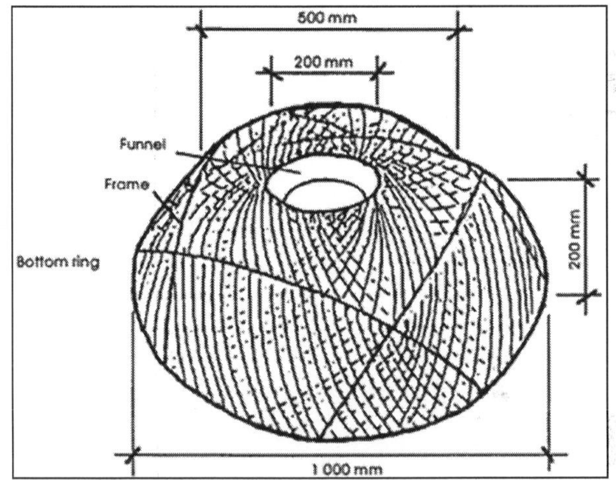

Fig. Experimental deep sea prawn pot (Thailand).

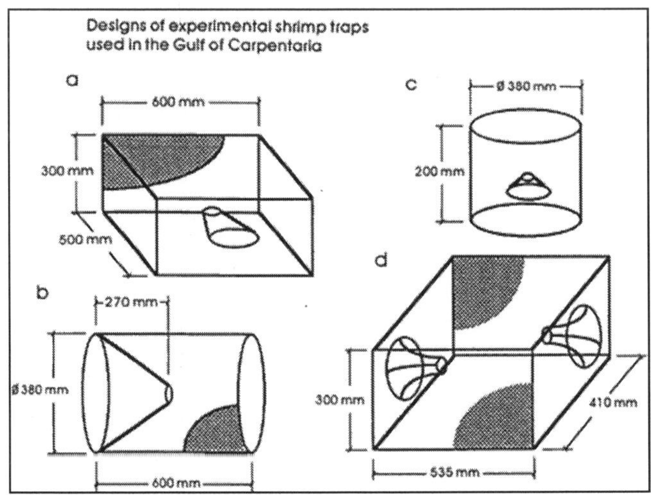

Fig. Australian experimental shrimp trap.

Lobster and rock lobster pots. There are two main types of lobsters: temperate clawed lobsters (Homarus), and tropical and subtropical rock lobsters (Panilurus) which do not have claws.

In Canada and North America, lobsters are taken principally with pots that have wooden frames, although there is a recent move towards the use of plastic-coated metal. Canadian lobster pots are usually made from timber and are covered with 80 mm diamond mesh made from PARtex 1030 twine. The two funnels are also made from this mesh. The details of two common pots used in Canada.

Similar pots are used in North America and Europe. Beehive pots are also commonly used. The two most common types of rock lobster pots are the beehive and the slat. Both rectangular and "D"-shaped wooden slat pots and beehive pots are used in the western Australian rock lobster fishery; a common design for a rectangular slat pot. The beehive pots used in this and the southern and eastern Australian rock lobster fisheries are of similar design to those used in other parts of the world. In eastern Australia, "D"-shaped pots covered with wire mesh are also used. In New Zealand, rock lobster traps are made from welded mesh and require no internal frames.

The rock lobster fishery on the northeast coast of Brazil traditionally uses pots made from mangrove branches covered with chicken wire or synthetic mesh. The details of these pots (covos). In Nicaragua, rock lobsters are fished with pots made from 50 × 25 mm galvanized steel mesh with a funnel woven from cane or bamboo.

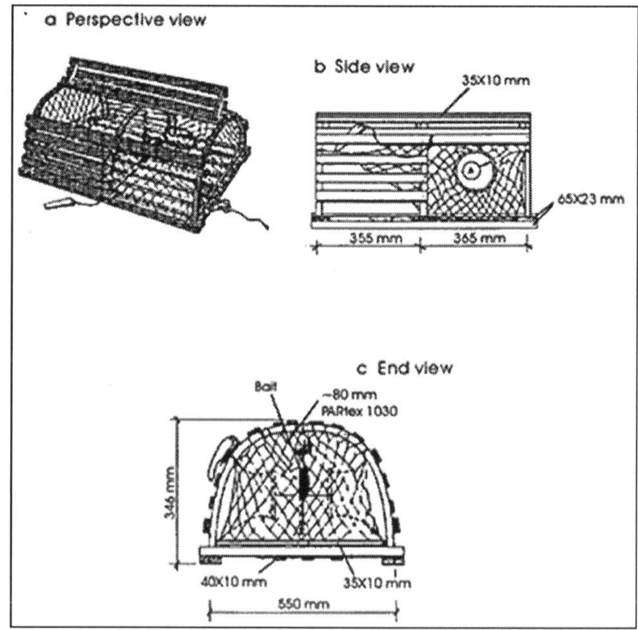

Fig. Lobster pot (east coast of Canada).

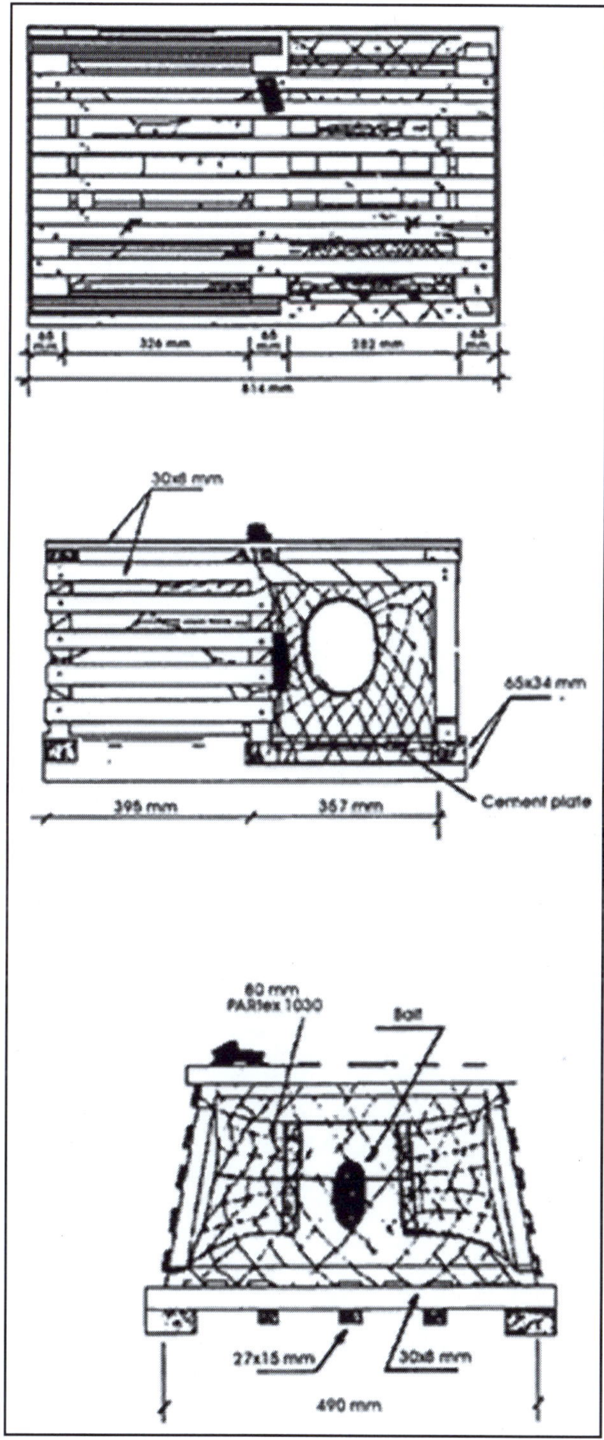

Fig. Lobster pot (east coast of Canada).

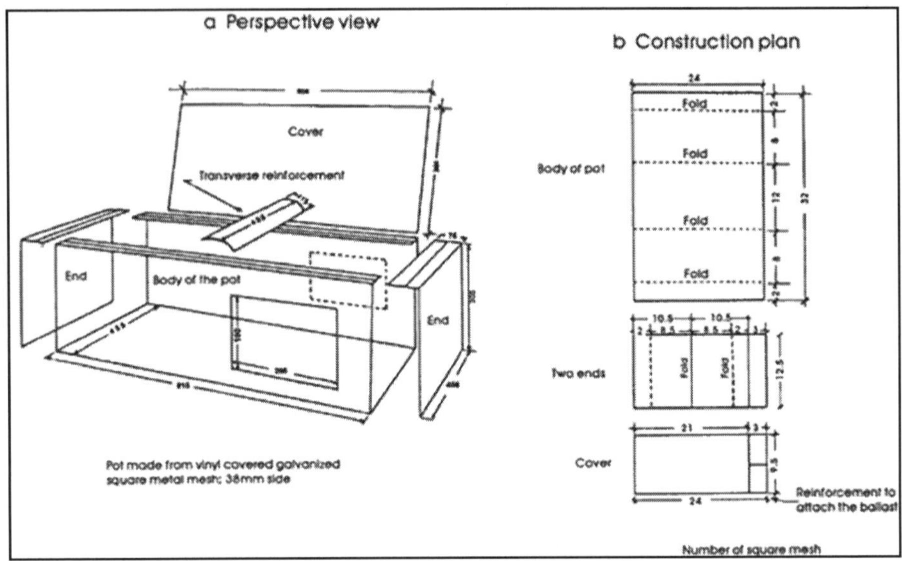

Fig. Canadian plastic-covered metal lobster pot.

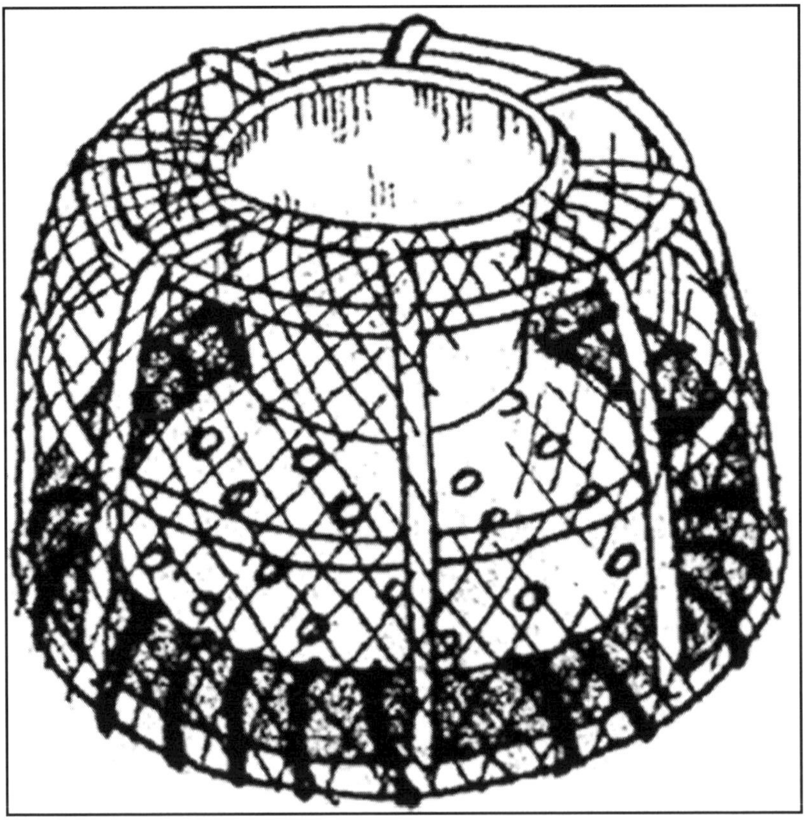

Fig. Lobster pot in plastic tube (Brittany, France).

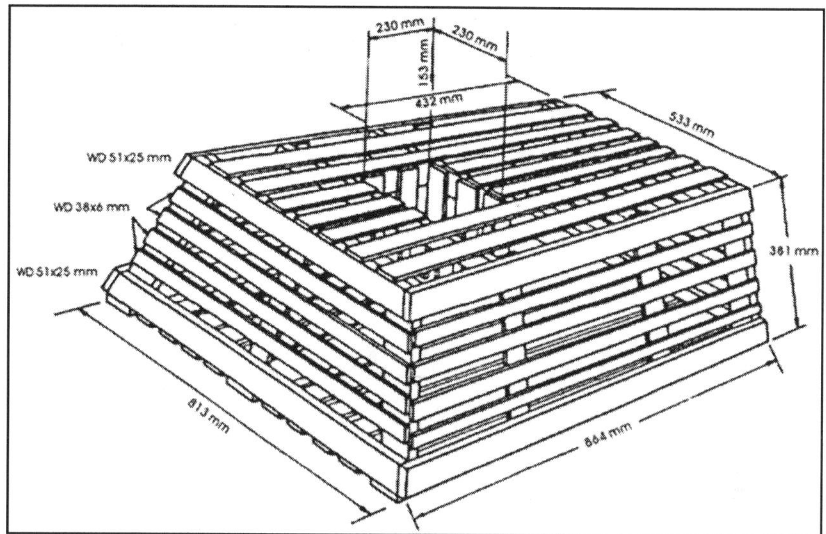

Fig. Western Australian batten rock lobster pot.

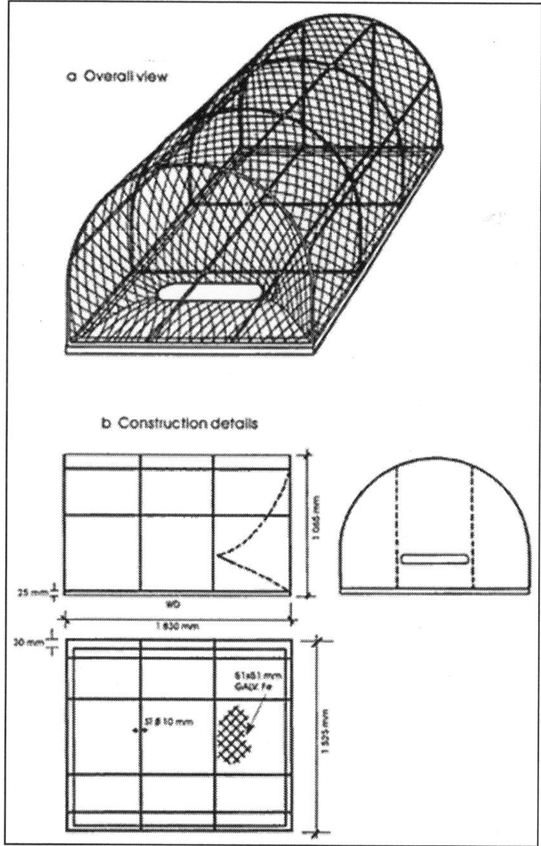

Fig. Eastern Australian wire netting rock lobster pot.

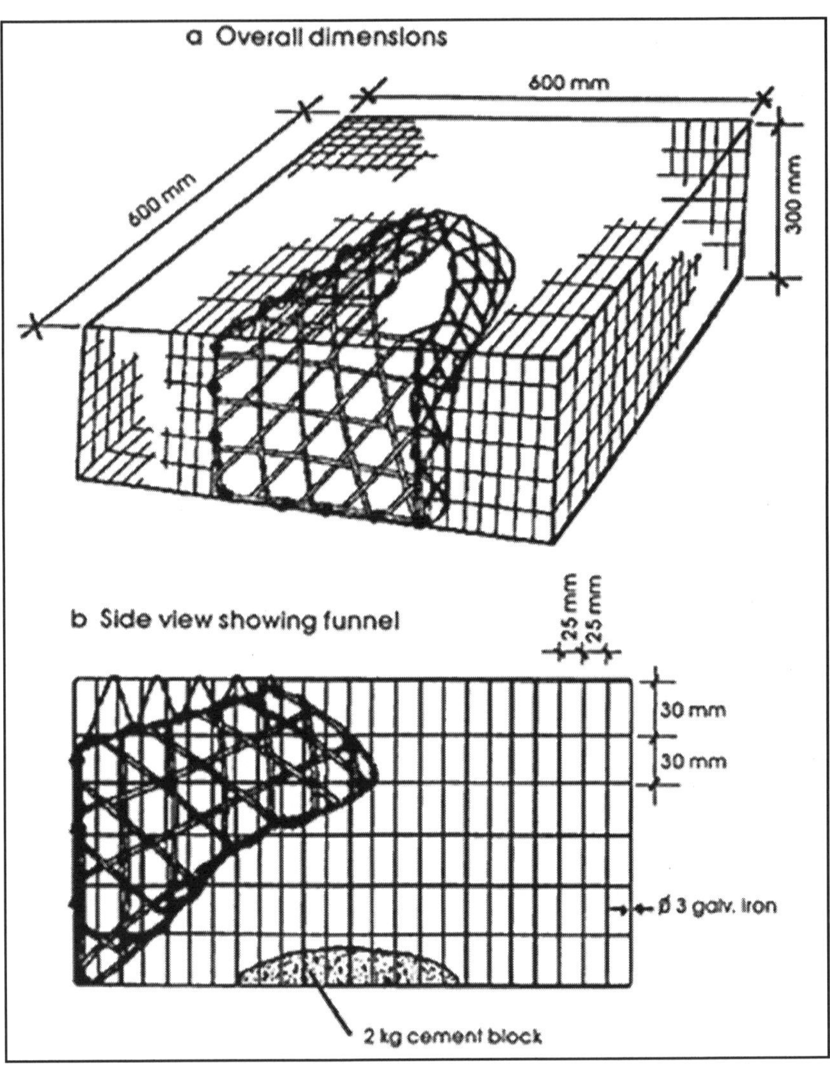

Fig. Rock lobster pots from Nicaragua.

King crab traps. Deep-water fishery for king crab in the northeastern Pacific and the eastern Bering Sea is carried out with large pots constructed from galvanized steel with frames of 213.3 × 213.3 × 91.4 cm. The frames for the funnels are also made from galvanized steel. The pot is covered with 152 to 229 mm polypropylene mesh and the funnels are covered with 76 to 114 mm mesh.

Swimmer crab pots. Estuarine and oceanic swimmer crabs provide an important fishery in some tropical and subtropical areas. Although the crabs are often taken by hand or with hand nets, potting is also an important fishing method. As crabs are generally not large and not very active swimmers, crab pots are small and simple in design. Crab pots can be shaped from wire netting or constructed from welded mesh. In some places a metal hoop with a strong synthetic mesh laced on to it is used. Such pots (sometimes called "dillies") are set with bait in the centre and hauled quickly so that the crabs cannot escape during hauling. In northern Australia, mud crabs (Scylla serrata are taken with a specially made trap, which is known locally as a pot. These are typically 600 × 700 × 200 mm and made from 50 × 75 mm galvanized wire welded mesh without a supporting frame.

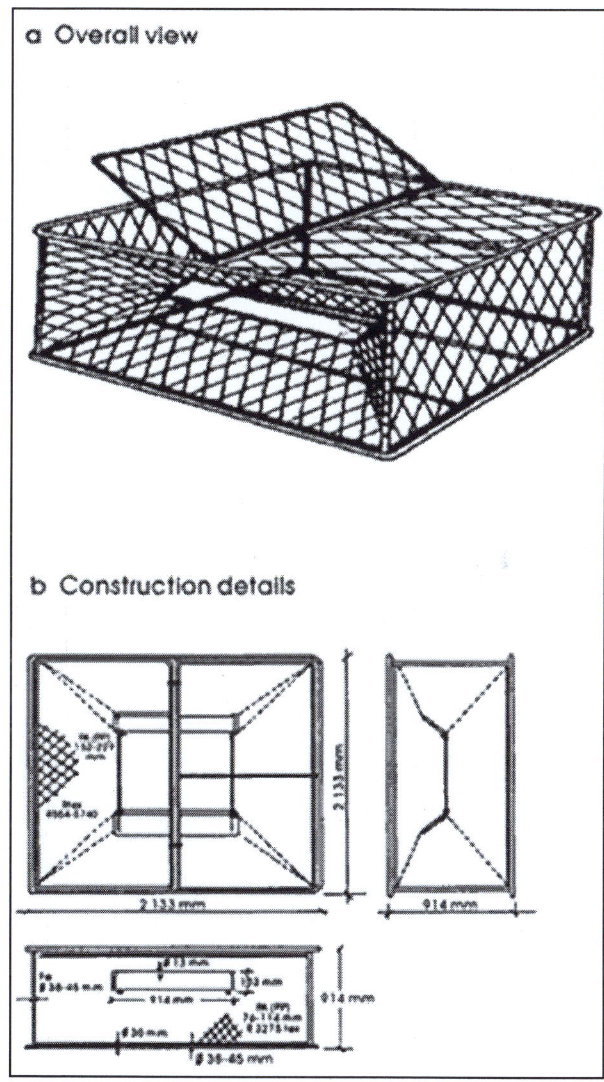

Fig. King crab pot.

To make the pots, a rectangular piece of mesh 1 800 mm long by 600 mm wide is cut for the top, bottom and sides of the traps. Two additional pieces of 700 × 200 mm are then cut to make the end of the trap. When cutting the mesh, leave the ends of each wire long, so that they can be twisted to secure the top and ends of the pot. The funnels are made as a straight oval from 15 × 15 mm square plastic mesh (sometimes known as "gutter mesh"), the oval being 225 mm wide by 100 mm deep, i.e. 3 × 2 meshes in the wire.

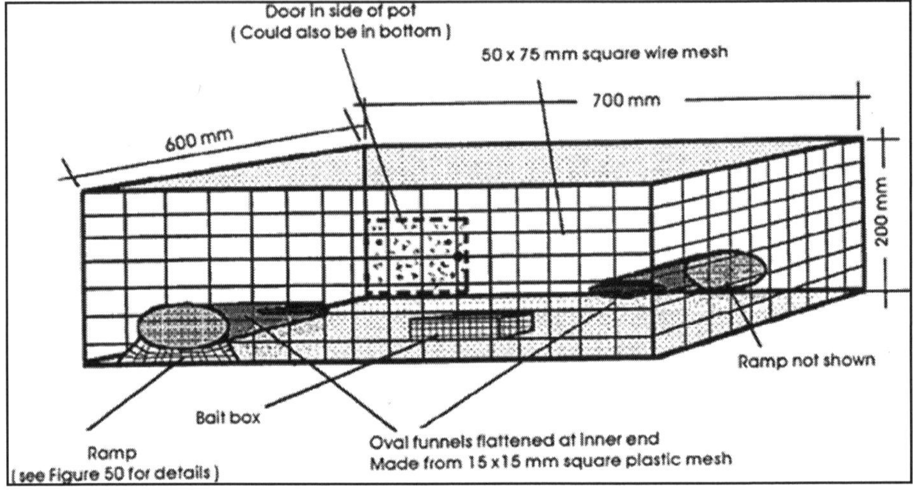

Fig. Northern Territory (Australia) mud crab pot.

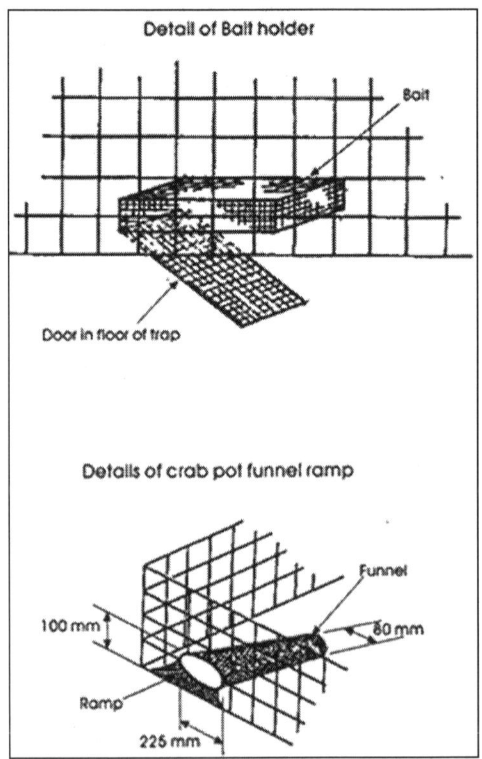

Fig. Bait holder and funnel ramp for crab pot.

As crabs' bodies are wider than they are high, these funnels are flattened to approximately 80 mm high at the inner end. Sometimes a ramp is placed up to each funnel to make it easier for the crabs to walk up. This ramp also places the funnel entrance above the crabs when they are inside. This shape will also discourage the entry of unwanted fish. A 300 × 200 mm door is placed in the back of the trap to make emptying easier. Access to the bait boxes or bag can be made easier by placing an opening on the bottom of the pot. Depending on the tidal range where they are set, crab pots are rigged with approximately 7 m of rope and 10 cm (4 inch) buoys to mark their location.

OCTOPUS, SQUID AND CUTTLEFISH POTS AND TRAPS

Although most squid and cuttlefish are taken with fishing gear other than pots and traps, some quite effective simple traps are used in several parts of the world. Octopuses are traditionally caught with traps as well as by hand.

Octopus traps. Unglazed earthenware pots are used in traditional octopus fishing in the Mediterranean and southeast Asia. The pots are placed on the sea bed for octopus to use as shelter. They differ in size and shape depending on the size, type and behaviour of the octopus being targeted. Although earthenware pots are traditional, you can also use other materials such as plastic piping, steel piping, old motor tyres and large empty mollusc shells to make traps for octopus, as they will enter almost any receptacle that can provide shelter.

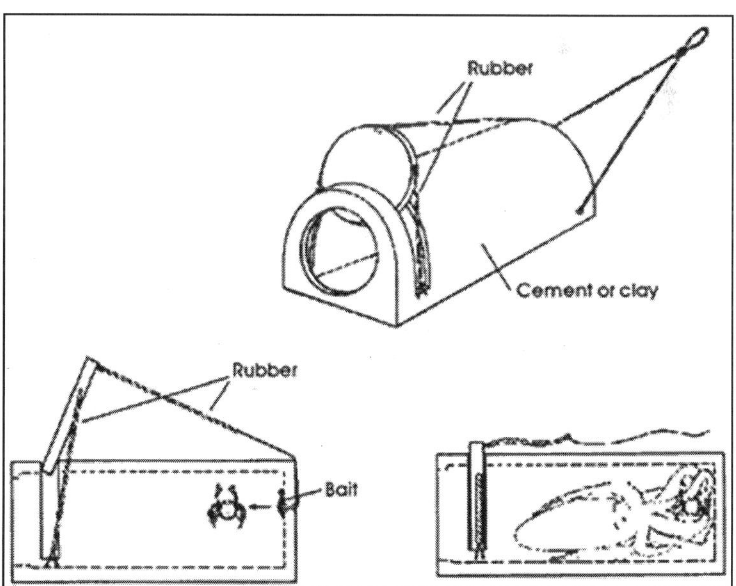

Fig. Baited octopus trap from Japan.

Across the world, both baited and unbaited octopus and cuttlefish pots and traps are used. An example of a baited trap. This Japanese trap is made from concrete or clay and has a door of stretched rubber that springs across its entrance when the octopus pulls on the bait. A type of unbaited trap used in Venezuela and Japan.

Another kind of octopus trap can be made from two 50 cm long pieces of 150 to 200 mm diameter PVC piping lashed or bolted together lengthways. A flat concrete block is bolted into the centre of each pipe. These traps can easily be modified to include a door at each end of the tubes. The doors are connected to a bait inside the PVC pipe in such a way that when an octopus enters the trap and pulls the bait, the door closes behind it.

Pairs of tubes are attached to a buoyed main line and set in lines on the bottom. Old tyres cut into pieces and split lengthways can also be used to make effective octopus traps. A way to make these traps. It has been reported that the rubber on the outside of the tyres peels off over time but, as old tyres cost little, they can be replaced regularly. These traps are set in the same way as the PVC piping traps and the earthenware pots. Octopus pots are normally set overnight attached to longlines with up to 100 pots on each line. They are hauled back up in the morning. The number of traps on each line depends on the extent of the grounds and the size of your vessel.

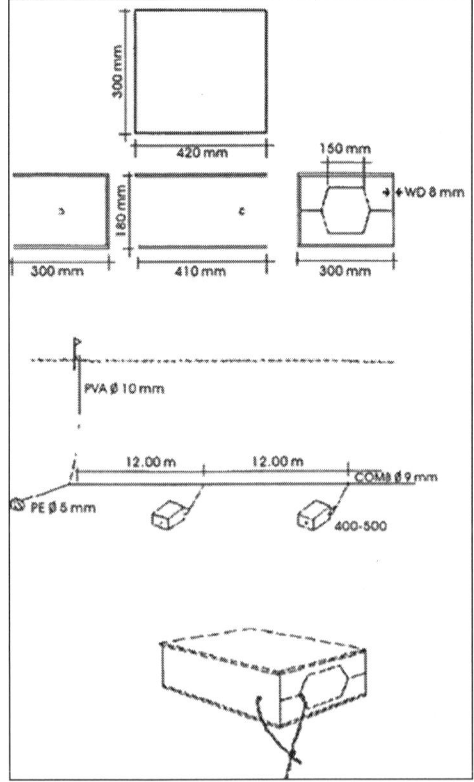

Fig. Unbaited octopus pot from Japan and Venezuela.

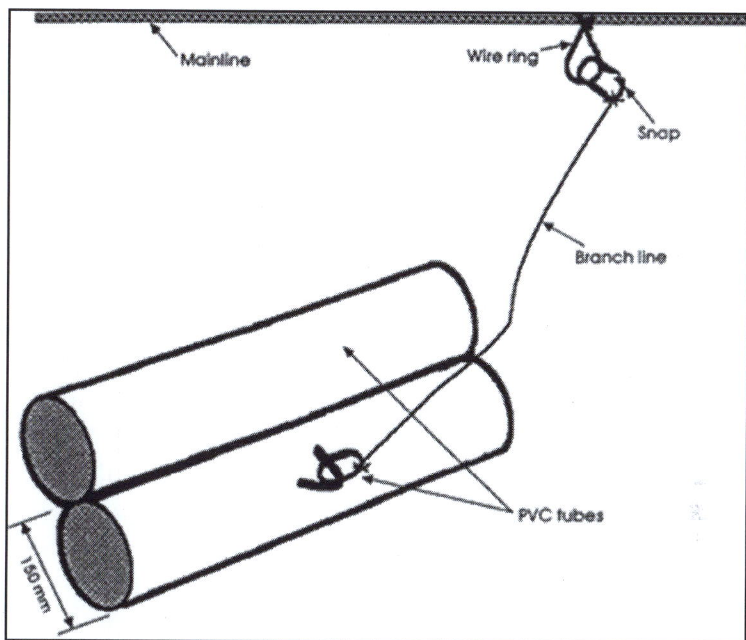

Fig. Unbaited PVC pipe trap for octopus.

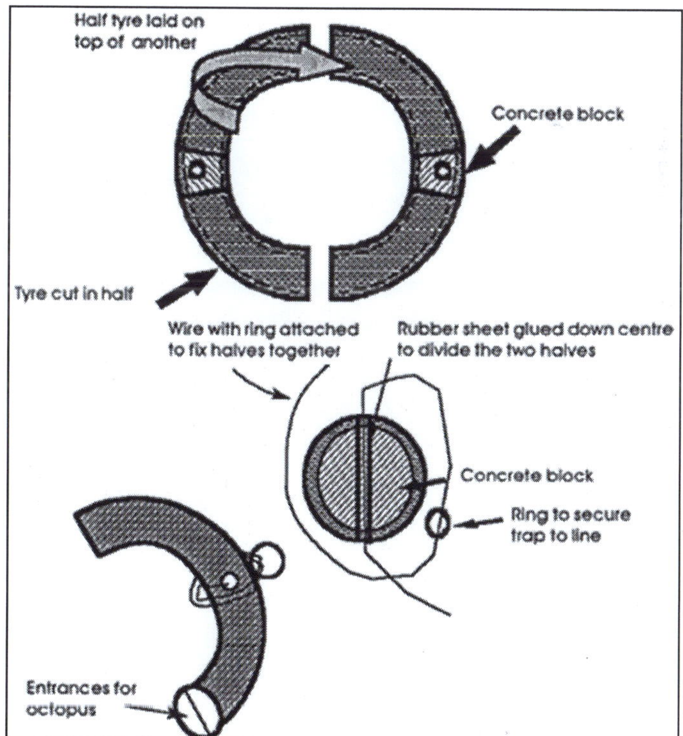

Fig. Octopus trap made from old tyre.

Squid and cuttlefish traps. Several types of squid and cuttlefish traps are used in Southeast Asian countries. A rectangular Indonesian trap is described in detail in this manual.

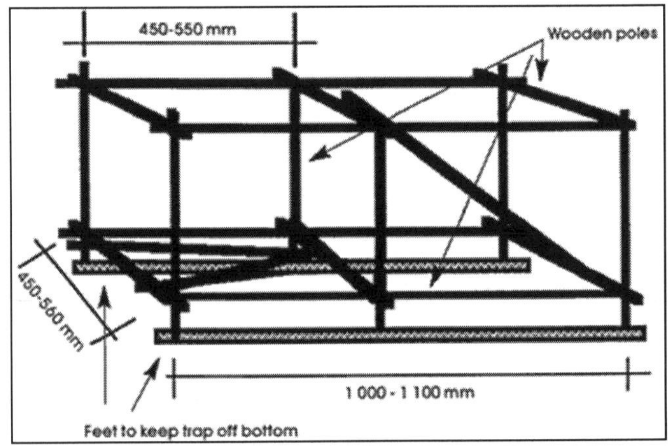

Fig. Wooden frame for an Indonesian trap for squid.

This rectangular trap is 100 to 110 cm long and 45 to 56 cm square at the ends. The frame is made from small wooden poles lashed together. This is then covered with PA 210/18 4.4 cm mesh.

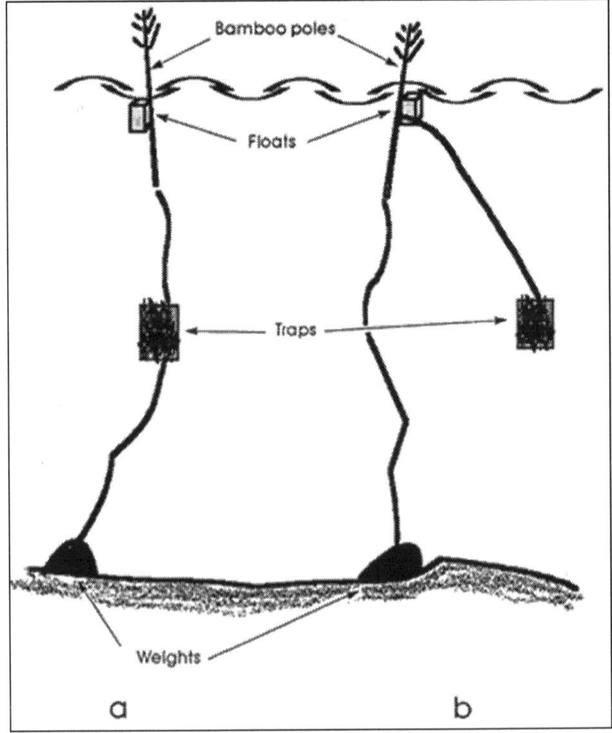

Fig. Rigging of cephalopode Indonesian trap.

The funnel follows the two inside poles and is no wider that 35 cm at the inner end. When the trap has been completed, it is covered on the top and bottom with coconut or similar leaves. The traps are rigged either on a single line or on a double or side line. Squid eggs are used as bait for both squid and cuttlefish, but sometimes white plastic bags or broken white china are used instead. The trap should be set about 2 to 3 m from the sea bed, taking care not to allow it to touch the bottom.

POTS AND TRAPS

Pots are considered within the International Standard Statistical Classification of Fishing Gear to be a type of trap (Nédélec and Prado 1990) but are described separately here because of the differences in catching principle and construction between pots and other forms of trap. The general catching principle of pots (creels) and traps is to entice or lead the target species into a box or compartment from which it is difficult or impossible to escape.

POTS

Catching Principle

As with longlining, pot fishing is normally based on attracting target organisms by bait (chemical stimuli). When attracted to the pot, the target organism must enter the pot to gain access to the bait. This can be done through one or several entrances (funnels) of the pot.

Construction

Typical pot shapes are box, cone, cylinder, sphere or bottle. The size of pots may vary from small crayfish pots (conical: 0.3 m diameter and 0.2 m height) to large king crab pots (box shaped: 2×2×1 m). The pot entrances are usually funnel- or wedge-shaped so that the target organism is led into the pot fairly easily, but with low probability of escaping. Pots may be constructed from various materials like wood, palm leaves, metal frames lined with webbing, wire mesh or plastic materials.

Operation

Pots are normally set on the bottom, either as single pots with a buoy line to the surface or in strings of several pots connected to a main line at certain intervals. Pot gear is usually soaked overnight, but longer soak times may be used in certain fisheries. The operation cycle is similar to that of longlining, with baiting, setting, fishing and retrieval. The bait is either freely suspended in the middle of the pot, or put in perforated bait containers to prevent it from being eaten by scavengers. As in longlining, different pelagic species like sardines, herring and mackerel are typically used for pot bait, but most kinds of fish and mussels etc. may be used.

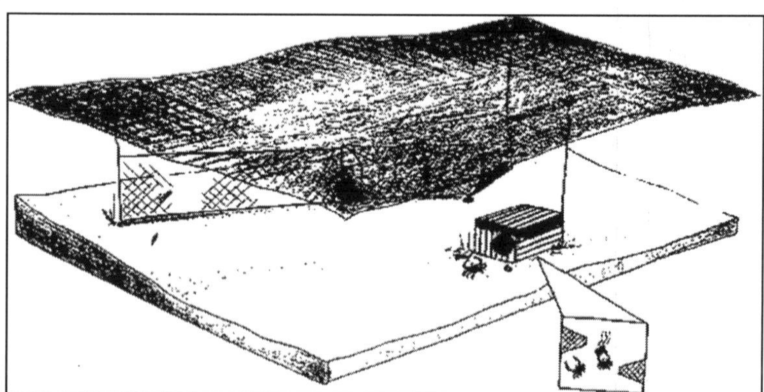

Fig. Catching principle and construction of pots (lower) and traps (fyke net - top). Expanded view: schematic illustration of entrances (funnels) and bait.

Target Species

Pots are most widely used to catch different crustaceans, like crabs, lobsters and shrimps. Pots are also used for catching different species of finfish like sablefish, tusk and cod in temperate waters and reef fish such as groupers in tropical waters. Other species that are caught with pots are whelks and octopus.

TRAPS

Catching Principle

Traps are normally not baited, but catch fish and other organisms by leading them into the trap, eventually to the fish compartment, that is designed for holding the fish entrapped with low possibility of escaping.

Construction

Compared with pots, traps are usually larger and often more permanent constructions. Tidal traps are based on walls or fences forming V-shaped constructions that entrap fish that have come in with the tide as the tide goes out. Typical salmon and cod traps are cage-like constructions made from webbing with long leader nets to guide the migrating fish into the trap. Tuna may also be caught with traps of this type. The fyke net is a smaller form of trap, with a leader net connected to the trap or "fyke" which usually consists of three compartments with funnels leading from the outer to the middle and finally to the inner compartment or "fish bag".

In tidal traps the fish are confined sideways by the leader walls, above by the sea surface and below by the sea bed. In floating fish traps like those used for cod and salmon, the fish are confined by side and bottom net panels while the sea surface serves as the top confinement. Fyke nets are set under the surface and thus the fyke (trap) has to be completely lined with webbing to confine the fish.

Operation

Tidal traps are usually permanent constructions where the fish compartment is emptied at low tide. Cod and salmon type traps are usually set out for the season and operated for one to a few months, emptying the fish compartment on a daily basis. Fyke nets are operated like pots, set one by one in the littoral zone, from one to ten meters depth, and are usually moved to another spot after retrieval, usually on a daily basis.

Target species

A variety of target species are caught by tidal traps, both finfish and crustaceans, e.g. shrimps, naturally dominated by species living in the tidal zone. As mentioned above, traps are traditionally used for catching cod and salmon (N. Atlantic), tuna (Mediterranean), small pelagic species in Far East Asia, some species of weakfish (members of the Sciaenidae) and others. Fyke nets are used for catching various species, but are particularly used for eel and cod.

ACTIVE FISHING GEARS

Fish capture by active gears is based on the aimed chase of the target species and combined with different ways of catching it.

SPEARS AND HARPOONS

This is one of the most ancient ways of active fish capture.

The catching principle and construction of spears and harpoons.

Catching Principle

Capture with spears and harpoons depends on visual observation of the target species, which is then impaled by the spear or harpoon from a relatively short distance.

Construction

Basically, the spear or harpoon is designed for easy penetration of the target organism, but the spear head is equipped with barbs or flukes that hold the prey when it is hit. Usually the spear or harpoon is connected to the fisher and boat by a line, so that it can be retrieved, with or without catch.

Operation

Spears and harpoons are most often operated from a vessel, but can also be used from land.

Target Species

Common target species with this fishing method are flatfish, swordfish, tunas and whales.

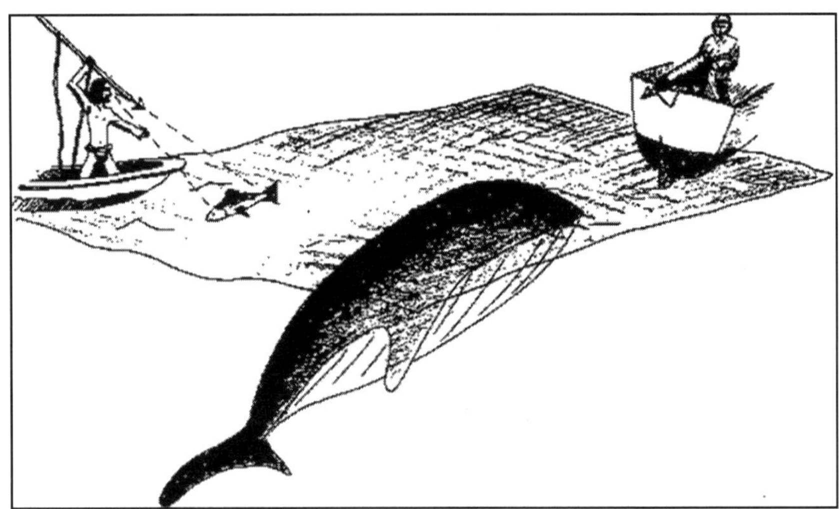

Fig. Catching principle and construction of spears and harpoons

TRAWLS AND DREDGES

Trawls and dredges are often called towed gear or dragged gear.

Catching Principle

The catching principle and construction of trawls.

Construction

Trawls and dredges are in principle netting bags that are towed through the water to catch different target species in their path. During fishing, the trawl entrance or trawl opening must be kept open. With beam trawls and dredges this is done by mounting the trawl bag on a rigid frame or beam. With otter trawls the opening is maintained by so-called otter boards (trawl doors) in front of the trawl which keep the trawl open sideways while the vertical opening is maintained by weights on the lower part (ground-rope) and floats on the upper part (headline).

With pair trawling, the vertical opening is also maintained by weights and floats, while the lateral opening is maintained by the distance between the two vessels that are towing the trawl. In otter trawling, the trawl is connected to the trawl boards by a pair of sweeps (rope or steel wire) and the trawl doors are connected to the vessel by a pair of warps (normally steel wire). In otter trawling and partially in pair trawling, the sweeps and warps are also part of the catching system, as they will herd fish towards the centre of the trawl path and the approaching trawl, so that the trawl may catch fish over a larger area than that of the trawl opening. With beam trawl and dredges there is little or no herding of target species in front of the trawl, so the effective catching area is that of the trawl or dredge opening.

Fig. Catching principle and construction of an otter trawl, showing the trawl connected by the sweeps to the trawl doors (otter boards) and the warps between the trawl doors and the vessel.

Operation

Beam trawls and dredges are exclusively operated on the bottom, where they are towed for a certain length of time (towing time) and distance before being retrieved for the emptying of the catch and being set again for another tow.

Otter trawls and pair trawls are most often operated on the bottom to catch different demersal target species. However, these gears are also commonly used for pelagic (or mid-water) trawling at different depths between the surface and the sea bed. This is done by attaching more floats to the head rope of the trawl opening as well as regulating the trawl depth by varying the length of warp and towing speed. In most pelagic trawling, the trawl depth is monitored by depth sensors on the trawl, so that the fishing depth can easily be adjusted to that of the fish targets.

Target Species

Beam trawls are mainly used for catching flatfishes such as plaice and sole as well as for different species of shrimp. Dredges are commonly used for harvesting scallops, clams and mussels. Demersal otter and pair trawls are used to catch a great variety of target species like cod, haddock, hake, sandeel, flatfish, weakfish, croakers as well as shrimps. Pelagic trawls are used in the fisheries for various pelagic target species, like herring, mackerel, horse-mackerel, blue whiting and pollock.

SEINE NETS

Catching Principle

Seine netting (including two variations known as Danish seining and Scottish seining) can be described as a combination of trawling and seining. When setting the gear, the first warp (rope) is attached to an anchor with a surface buoy (Danish seining) or a buoy only (Scottish seining) and set in a semicircle. Then the seine bag is set before paying out the second warp in another semicircle back to the buoy (attached to the anchor in Danish seining). When the seine and warps have sunk to the bottom, the warps are hauled. As they are tightened, the warps move inwards towards the centre line between the vessel and the seine bag. Fish in the encircled area will then be herded towards the central part of the area. As the warps are further tightened, the seine bag moves forward and catches the fish.

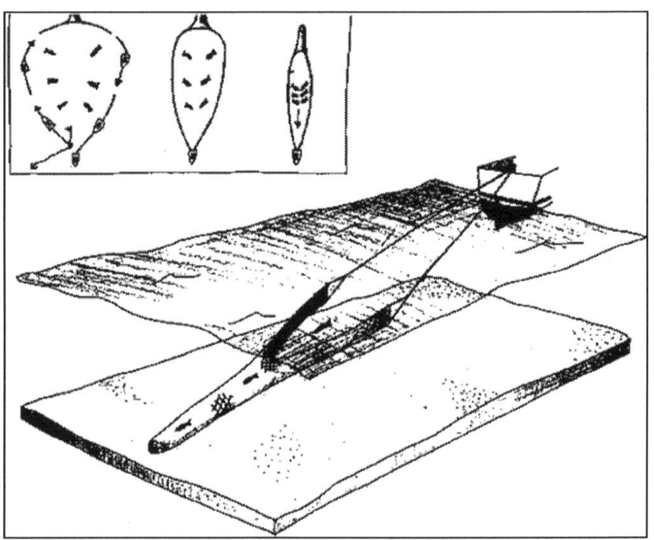

Fig. Catching principle and construction of seine nets. Expanded view: three stages of the catching process.

Construction

As mentioned above, the main parts of a seine net are the seine bag and the warps. The seine bag is similar to a trawl bag, where the entrance is kept open by floats on the headline and a weighted ground line (foot rope). The warps are usually made of heavy rope, so that they maintain good contact with the bottom for as long as possible during tightening in order to herd the fish towards the central area for later capture by the seine bag.

Operation

The seine net was originally constructed for the capture of flatfish on soft and smooth bottoms and operated as described above. In later years this gear has also been developed to be operated on rougher bottoms and in the pelagic zone. A more recent mode of operation is for instance used on mid-water shoals of cod. The fishing depth of the seine net is then determined by large surface floats connected to the head rope of the seine net by lines, the length of which corresponds to the desired fishing depth.

Target Species

The seine net is still commonly used to catch different flatfishes such as plaice and sole, but has in recent years become an important gear also for cod and other demersal target species.

BEACH SEINES

Catching Principle

The operation of beach seines is based on encircling fish schools by a netting wall, made of webbing where the meshes are so small that the target species does not get entangled.

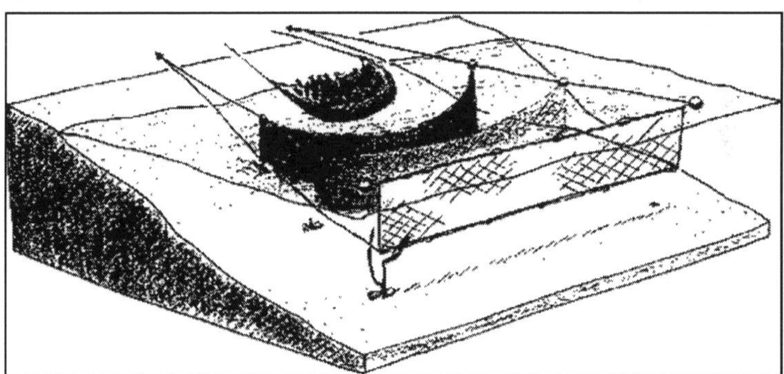

Fig. Catching principle and construction of (beach) seines, showing four stages of the catching process.

Construction

The beach seine is an ancient gear that is still widely used. The seine consists of a wall of webbing, e.g. with a depth of 5m by 100m length, with an upper float line and a lower sinker line. In principle a similar construction to the gillnet, but with smaller meshes so that the fish is entrapped instead of being gilled or entangled. At either end of the seine there are long warps (rope).

Operation

As the name indicates, the beach seine is operated from the beach - using the beach as an additional barrier in the catching process. The gear is normally operated from a small vessel. First, one of the end warps is paid out perpendicular to the beach. Then the seine is set parallel to the beach and the second end warp is taken back to the beach. The warps are pulled in so that the seine approaches the beach in a semicircular form - and most of the fish in the area between the seine and the beach are likely to be caught. In many seine fisheries, both beach seine and purse seine, light is used to attract and concentrate fish before the seine is set.

Target Species

Beach seines catch a variety of inshore fish species, both demersal and pelagic.

PURSE SEINES

Catching Principle

The purse seine is used to encircle fish schools in mid-water, close to the surface, by a netting wall with small meshes. The lower part of the net is then closed to prevent escapement by diving.

Construction

The purse seine was developed in the 20th century, for offshore fishing. Basically its construction is similar to that of the beach seine. However, below the sinker line, the purse seine is equipped with a series of metal purse rings spaced at regular intervals. By hauling in the purse line that runs through the purse rings, it is possible to purse and close the bottom part of the seine so that the encircled fish cannot escape.

Operation

The purse seine is always operated from a vessel, varying in size from small coastal purse seiners of 15m in length to large ocean going purse seiners as large as 100m length. When a fish school has been located, the catching operation starts by dropping a surface buoy with a line connected to the end of the seine. As the vessel moves forward, the drag from the buoy line will pull the purse seine overboard and the seine is paid out in a circle around the fish school. When the setting is completed, the buoy is picked up, and the purse line is pulled in to purse and close the bottom of the seine. Then the seine is pulled in until the fish are concentrated in the last (and often reinforced) part of the seine, from where it is taken on board with a brailer (large dip net) or by a fish pump.

In modern purse seine fisheries, hydro-acoustic equipment (sonar) is widely used for locating fish schools and also for monitoring the position of the school relative to the gear during the setting of the seine.

Target Species

Purse seine fishing is used almost exclusively for pelagic fish like herrings, sardines, sardinellas, anchovies, mackerels and tunas.

OTHER FISHING GEARS AND DEVICES

The most common fishing gears and methods are briefly described above. There are, however a great variety of different gears which are specialized varieties of the main gear types and methods. It is recommended to use some of the more comprehensive works on fishing gears and methods. Some devices and techniques are, however, used with different fishing gears to improve the catch efficiency:

Light is used to attract fish in many fisheries, but is most often used with purse seining, beach seining or other varieties of seine fishing. In darkness the light will either attract the target species directly or indirectly by attracting and illuminating prey organisms. The light source is often lamps mounted on a vessel or small raft. After some time the net is set around the light source with the attracted fish.

FADs (fish aggregating devices) are also commonly used in some areas to aggregate fish. These can consist of anchored rafts of logs or other material and act as artificial habitats that will attract fish and other organisms over time and hence create good fishing spots that can be exploited by different gears. This is also the case with artificial reefs that either are made by shipwrecks or by deliberately dumping objects on otherwise flat and sandy sea beds to create fish habitats. The Code of Conduct encourages the use of such structures providing they are used in a responsible manner.

Different stupefying devices are also used to capture fish. Explosives (dynamite) will stupefy fish, some of which will float to the surface so that they can be picked up. The use of explosives with fishing is, however, regarded as a very destructive practise as the explosion most often kills much more fish than those that are caught - and can in addition ruin valuable fish habitats like coral reefs.

Different chemicals can also be used in the same way. Rotenone (a poison derived from plants) is one of the best known examples used to stupefy fish, mainly in fresh water systems. As with explosives, the use of chemicals gives a high risk of killing much more fish and other organisms than those that are harvested. The use of chemicals with fishing should therefore be considered as not responsible. The Code of Conduct specifically calls for the prohibition of "dynamiting, poisoning and other comparable destructive fishing practices".

GEAR SELECTIVITY AND ECOSYSTEM EFFECTS OF FISHING

Before commenting on the selectivity properties and ecosystem effects of different fishing gears it may be useful to give a brief description of relevant factors and definitions.

THE CATCHING PROCESS

The catching process starts as the fishing gear is deployed in the water and ends as it is retrieved from the water, be it ashore or on the deck of a fishing vessel. Throughout the catching process there may be encounters between the gear and various fish and other marine organisms, including sea birds, and bottom habitats.

ECOSYSTEM EFFECTS OF FISHING

The effect of fishing on the ecosystem is primarily the removal of the organisms caught in the fishery, but also includes the direct and indirect effects caused by the gear during the catching process - like destruction of bottom habitats (e.g. corals), "ghost fishing" by lost fishing gears, pollution etc.

SELECTIVITY

The selectivity of a certain fishing method depends on its ability to select the desired ("target") species and sizes of fish from the variety of organisms present in the area where the fishery is conducted.

The total selectivity of a fishing method is the combined result of the inherent selective properties of the fishing gear and the way it is operated. With most fishing gears it is possible to impair or improve the selectivity by changing the gear configuration or the operation. For example, in trawl fishing the catch of small fish can be reduced by increasing the mesh size and/or by the use of sorting devices like sorting grids or large mesh panels that allow for escapement of the smaller fish. The fisher can also select for target species and sizes by avoiding areas and periods where there is a high probability of catching small fish or otherwise undesired bycatch. The Code of Conduct requires the minimisation of the catch of non-target species and of discards.

BYCATCH

Bycatch is anything that is caught in the fishing process beyond the species and sizes of the targeted marine organisms. There is a great variety of bycatch species, ranging from sponges and corals to unwanted or unmarketable fish species or sizes, as well as turtles, marine mammals and sea birds. Bycatch can be classified into three main groups: marketable and legal; non-marketable; and/or non-legal. Non-economic bycatches consist of organisms that are non-marketable for the fisher, while non-legal bycatches are sizes or species of marine organisms that are protected by regulations.

Thus, marketable and legal bycatch is welcomed by the fisher, while all other bycatch should be avoided. However, the capture of some bycatch is normally unavoidable. Most fisheries regulations do therefore allow for a certain amount of bycatch, e.g. a certain percentage of undersized catch of a target species or a certain amount of an otherwise protected species.

For example, in the New Zealand trawl fishery for squid, a certain number of sea lions are tolerated as bycatch, but the fishery is closed as soon as this number is reached for the specific fishing season. In the Barents Sea cod fishery, a bycatch of 15% (in numbers) of undersized fish (less than 42 cm) is tolerated within the legal frames of the fishing regulations.

DISCARDS

Discarding, which means to throw parts of the catch back into the water is a common practice in most fisheries, although the amount of discard varies significantly between different fisheries. Discards are most often organisms that are not marketable or give a low price compared with the more valued target species. The survival of discarded organisms depends on the ability to survive in air, the time they are kept out of the water and how they are handled before being discarded. However, it should be expected that most discarded organisms suffer high mortality, which adds a "hidden mortality" to the fishing mortality that is calculated from the landed catch.

Unintentionally, modern fisheries management has encouraged increased discarding in many fisheries. With the introduction of quotas and licences for different species it is often illegal to catch and land certain species of fish. Thus it is not uncommon, particularly in mixed-species fisheries, to find large-scale discarding of valuable, marketable fish (e.g. cod and saithe) because the trip quota or monthly quota has been exceeded. In addition, in some fisheries, small fish under the legal size limit for landings may also be discarded.

"High grading" is another form of discarding where only the most profitable part of the catch is retained, while less valuable fish are discarded. This phenomenon is also often linked to a quota system where the fisher tries to get the maximum value from a limited quota by keeping only the most valuable part of the catch and discarding the rest.

BY-MORTALITY

By-mortality is the mortality of marine organisms from injures caused by encounter with the fishing gear during the fishing process. One example is fish that die from infections or osmotic imbalance caused by scale loss after escapement through trawl or gillnet meshes. The introduction of mesh size regulations e.g. in trawls should therefore take into account, and be accompanied by studies of, the survival of fish that are released through trawl meshes or sorting devices. If the released fish suffer high mortality there is little to be gained by sorting it out of the fishing gear, which is the case with vulnerable species like herring. On the other hand, studies have shown that cod and several other demersal species have high survival rate after escapement from or encounter with fishing gear.

GHOST FISHING

The term "ghost fishing" is used to describe the capture of marine organisms by lost or abandoned fishing gear. This is particularly a problem with gillnets, trammel nets and pots. The gear is usually lost because it becomes stuck on rough bottoms containing corals and stones, causing the buoy line to break during retrieval. Nets or pots may then continue to fish for years.

HABITAT EFFECTS

Destruction of bottom habitats is particularly a problem with the use of dragged demersal gear like beam trawls, otter trawls and dredges. Corals and other epifauna have been and may be destroyed over large areas. It is still debated as to whether these gears have any real negative effect on soft, sandy bottoms.

However, it has been documented that trawling has ruined large areas of coral, which has a very low recovery rate, and other epifaunal organisms. Here, the Code of Conduct calls for the development and use of environmentally safe fishing gear.

CATCH QUALITY

Properties of fishing gears and the way in which they are operated also affect the quality of the catch, thus having an indirect ecosystem effect by the misuse of natural resources. In gillnet fishing, poor quality results from too long a soak time.

This results in fish dying in the nets and either rotting or becoming damaged by scavengers; therefore, that part of the catch is not marketable and has to be discarded. This may also be a problem in longline and pot fishing. In trawling, particularly with large catches, it is not uncommon that part of the catch is ruined by squeezing in the trawl bag or becomes of inferior quality because of too long storage on deck before it is processed. This too is contrary to the requirements of the Code of Conduct.

ENERGY EFFICIENCY

Use of energy, particularly fossil fuel, is also an ecosystem related aspect of fisheries. The energy efficiency (i.e. fuel consumption per unit of landed catch) varies considerably with different fishing gears and methods, from negligible use of fuel to more than 1 litre of fuel per kilogram of landed catch. This is covered in Sub-article 8.6 of the Code of Conduct, calling for energy optimization.

POLLUTION

Fisheries can contribute to air pollution through emission of combustion gases. The relative pollution effect from different fisheries is closely related to their energy efficiency. Pollution of water from fisheries is mainly by loss of fishing gear or by deliberately discarding old gear and equipment as well as oil products and chemicals at sea. These two aspects are covered by Sub-articles 8.8 and 8.7 respectively of the Code of Conduct.

SELECTIVITY PROPERTIES AND ECOSYSTEM EFFECTS OF DIFFERENT FISHING METHODS

The following is a description of the general selectivity properties and ecosystem effects of the different fishing gears and methods. For a proper evaluation of the ecosystem effects of fishing, each specific fishery has to be analysed separately, as the selectivity properties and ecosystem effects of a certain fishing method may vary considerably with geographical area, time of year and how the gear is operated.

A few examples are the following:

- Gillnets that are operated in shallow water and are hauled on a daily basis will yield catches of higher quality and with less risk of gear loss and ghost fishing compared with gillnets operated in deep water with a soak time of several days.
- Pelagic gillnets that are fished in the vicinity of sea bird breeding sites may have high bycatches of sea birds in the breeding season, but not in other periods.
- Bycatch of juvenile fish in demersal trawling may vary considerably with the availability of juveniles, the species composition, towing speed and general catch rates.

So even if the intrinsic selective properties and other ecosystem impacts of a certain fishing gear may be regarded as fairly constant, the effects caused by the gear on fish stocks and the rest of the ecosystem may vary with diurnal, seasonal and long term changes in species and size composition of organisms available to the fishing gear and with differences in fishing practice.

GILLNETS

In general gillnets are considered to be very size selective, with catches of fish sizes that correspond well to the chosen mesh size. However, due to entangling a small proportion of larger and smaller fish may be taken. The species selectivity of gillnets is not particularly good and as different fish species grow to different sizes, there is always a possibility of catching juveniles of a large species when using small mesh gillnets for a smaller target species. Another negative impact of gillnets is the bycatch of sea birds, marine mammals and turtles. Although little information exists on the real effect of such bycatches on the populations of these organisms, it has generated concerns, particularly for pelagic gillnet fishing.

Information on by-mortality of fish after escapement from gillnets is scarce. However, observations of fish with wounds from gillnet meshes are commonly made in catches by other gears, but the actual mortality rates from such injuries are not known. Ghost fishing is one of the most criticized aspects of gillnet fishing, and may have severe negative effects, particularly in deep water gillnet fisheries. The energy efficiency of gillnet fishing is generally high with a correspondingly low air pollution effect. As mentioned above, the catch quality of gillnet caught fish can be high. However, gillnets that are operated with soak times of several days tend to produce catches of inferior quality, as fish caught early in the fishing period may die and start to deteriorate long before the nets are retrieved.

TRAMMEL NETS

Compared with gillnets, trammel nets have very poor size selective properties and they will also catch a greater variety of species. The problem of ghost fishing is, however, lower as trammel nets generally are operated in shallow water with less risk of gear loss, but ghost fishing problems should nevertheless be anticipated due to loss of trammel nets that get stuck on rough bottoms like coral reefs.

HANDLINING AND TROLLING

Handlining and trolling are not particularly size selective and in principle not very species selective either. However, these gears are commonly used in specific seasons or at specific grounds where the fishers, by experience, are able to catch only one or a few species, so that the catches are usually dominated by a few targeted species. Otherwise, handlining and trolling are generally regarded as ecosystem-friendly ways of fishing which produce catches of high quality.

LONGLINING

Despite the fact that longlines may attract and catch a large variety of fish species and sizes, this gear is considered to have medium to good species and size selective properties. The species selectivity of longlines can clearly be affected by the type of bait used, as different species have been shown to have different bait preferences. The size selective properties can partly be regulated by the hook and bait size as many studies have shown a correlation between the size of hook and bait and the size of the fish caught. The longline attracts fish from several hundred meters away, and as large fish have a greater swimming and feeding range than smaller fish, this adds to the size selective properties of longlines.

Bycatch of marine mammals is no particular problem with longlining, but there might be significant bycatches of different seabirds, which mainly are caught as they try to catch the baited hooks during the setting of the lines. This problem has been recognised by FAO member States, leading to the development of the FAO International Plan of Action (IPOA) for Reducing Incidental Catch of Seabirds in Longline Fisheries. The IPOA specifies some optional technical and operational measures for reducing the incidental catch of seabirds, including, for example, increasing the sinking rate of baits and the use of bird scaring lines which are towed behind the vessel, above the longline being set. Little is known about the by-mortality of fish in longline fishing, but fish that are lost during retrieval of longlines do often suffer mortality. Ghost fishing may be regarded as a minor problem with longlining and this gear is not considered to cause significant adverse habitat effects. The energy efficiency of longlining is generally high, with typical energy coefficients from 0,1 to 0,3 (kilogram fuel per kilogram of landed catch), which is in the same range as that of gillnetting. Longline caught fish are in general of high quality, but as is the case for gillnetting, long soak times may lead to reduced catch quality mainly due to bottom scavengers (sea lice, hagfish) that may attack and eat parts of the hooked fish.

POTS

As with longlines, the species selectivity of pots may be regulated by the bait used. Lobster fishermen do for instance often use "sour" or rotten fish as bait to avoid catching crabs in their lobster pots. As with longlines, the attraction of fish and crustaceans to baited pots tends to attract the larger animals in the fished area. The size selectivity of pots may be further improved by the use of so called escape gaps, the size of which allows for escapement of smaller animals. By-mortality is not regarded as a problem with pot fishing and this gear has negligible effect on bottom habitats. There is, however, a certain risk of ghost fishing, as lost pots may continue to fish long after they are lost. This can be reduced by having certain parts of the pot made from a bio-degradable material. Furthermore, pot fishing is regarded to have a high energy efficiency and good to superior catch quality, as the catch normally remains alive and in good condition.

TRAPS

Traps are usually constructed of relatively fine meshed webbing to avoid the tangling of fish and other organisms, so their size and species selectivity is generally low. As the caught animals usually stay alive, and as traps are most often operated in shallow water, they allow for easy release and high survival of unwanted catch organisms. With responsible fishing practices, the actual selectivity properties of traps may therefore be good and the by-mortality is low. Traps in general have little adverse impact on bottom habitats, they do not create ghost fishing problems and the energy efficiency and catch quality of trap fishing are high.

SPEARS AND HARPOONS

The capture of fish and other animals by spears and harpoons is probably one of the most environmentally friendly fishing methods. As the target is identified before capture, the fisher can be very selective regarding both species and size of prey. Fishing with spear or harpoon may give some by-mortality of wounded animals that escape and, when used in reef areas, the use of spears can lead to damage of coral, but apart from these, there are no substantive adverse effects related to ghost fishing or habitat destruction and the energy efficiency and catch quality is high.

PELAGIC TRAWLS

Pelagic trawls generally have high species selectivity as they are commonly used for catching schooling pelagic fish that tend to occur in single-species aggregations. The size selectivity is poorer as the fish bag of the trawl is usually made from small mesh webbing to avoid meshing by smaller individuals. Successful trials have been done with sorting grids that effectively release the smallest fish (e.g. with trawling for mackerel). However, these have not been applied in practical fishing, as many pelagic fish seem to suffer high mortality after being released from the fishing gear - mainly caused by the loss of scales which easily leads to secondary infections and osmotic imbalance. At present, sorting systems for the release and protection of juvenile pelagic fish is therefore not recommended. By-mortality is hence a minor problem with pelagic trawling and this gear, naturally, does not have any ghost fishing or habitat destructive effects.

The fuel consumption of pelagic trawling can be high, but still the energy efficiency might be relatively good as large catches are often made during short time periods. The catch quality of pelagic trawls is also relatively high, although large catches may give some squeeze and pressure damage to the fish in the trawl.

DEMERSAL TRAWLS

Demersal trawls are used for the capture of a great variety of bottom fish and crustaceans (mainly shrimps and prawns) while dredges are used to harvest molluscs (clams and scallops).

FISH TRAWLS

Otter trawls are widely used for the capture of different demersal fish species - most often in so-called mixed-species fisheries. The size selectivity may to a certain degree be regulated by the cod-end mesh size. Ideally, a certain mesh size should allow for the release of all fish below a certain size. However, the mesh size selection of trawls may be hampered in many ways. With increasing catch load in the cod-end, the meshes tend to stretch and close, so that the effective mesh size is significantly reduced. Clogging (closing) of meshes by fish getting stuck in them is another common problem that leads to poorer selectivity. The choice of mesh size for a certain target species may not give ideal selection of other species with different growth characteristics, a problem that is related to all mixed-species fisheries. Considerable research and development effort has been spent in recent years to improve the size and species selectivity of trawl gear. Different solutions have been developed and implemented, like the sorting grid that is now used in many demersal trawl fisheries. The sorting grid which is now mandatory in the Barents Sea bottom trawl fishery for cod and other demersal species. Most of the juvenile fish will escape through the openings between the metal bars of the grid, while larger fish will be retained.

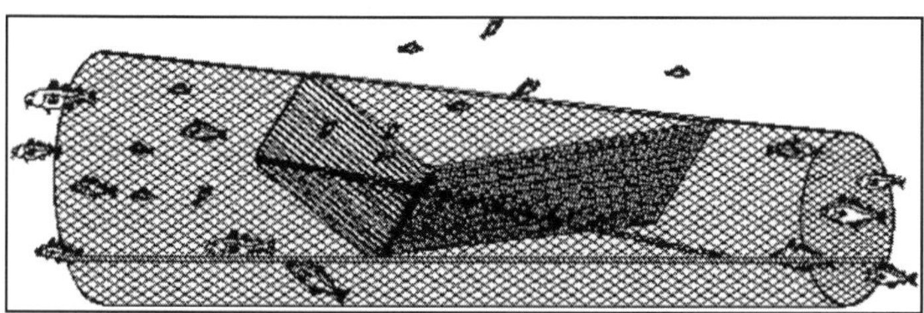

Fig. Section of fish trawl with sorting grid. Smaller fish are released through the grid slots, while larger fish swim or are swept down under the grid and go to the cod end.

Where introduced, sorting grids and other selective devices such as larger square mesh panels and escape gaps have lead to improved selectivity of bottom trawls, first of all the size selectivity, but also the species selectivity, although there is still a need for further improvements.

Releasing juvenile fish has little benefit if they do not survive. Extensive studies that have been carried out on demersal target species like cod and haddock have shown very low by-mortality of fish that have escaped through meshes or sorting grids. Although studies of survival after escapement so far have only been done for a restricted number of species, there seems to be a general indication of high survival of demersal fish after encounters with and escapement from fishing gears, when the fish are sorted out and released at fishing depth.

Demersal trawls do inevitably have an effect on bottom habitats. Several studies have been carried out within this field, however, without any general conclusions. On soft and sandy bottoms, there might be an inverse effect on the species composition in an area with a shift towards species that are less dependent on the epifauna removed by trawling. Most studies do, however, indicate that soft bottom habitats will be restored after some years without trawling

On hard bottom, trawling is likely to cause more long lasting or irreversible habitat effects, e.g. by destroying corals that have restoration periods from decades to more than a hundred years. Large areas of coral bed have already been destroyed by bottom trawling, particularly with the development of heavier and stronger trawl gear. Trawls are occasionally lost, but this gear loss is not associated with any risk of ghost fishing. The energy efficiency of demersal trawling is low and air pollution from the emission of exhaust gases is correspondingly high due to the high energy needed for pulling the net, doors, sweeps and warps through the water. The catch quality of trawl caught fish varies with the amount of catch and the towing time. Large catches do often lead to lower catch quality because of the squeezing of the fish in the trawl bag and a longer time before the last part of the catch is processed on board.

SHRIMP TRAWLING

Shrimp trawls are in principle comparable to demersal otter trawls for finfish capture and also have comparable ecosystem effects, except that the selective properties of shrimp trawling are very poor. This is due to the small meshes that have to be used in shrimp trawls in order to retain these relatively small target species. Shrimp trawling does therefore produce relatively large amounts of bycatch and a high proportion of this is discarded. The development of sorting grids has, however, improved the species- and size selectivity in many shrimp trawl fisheries, as most fish over a certain size are released from the trawl through the sorting grid or bycatch reduction device. Bycatch of the youngest fish groups (1-2 year olds) is still a problem, as they have overlapping sizes with those of the shrimps.

Bycatch of turtles during shrimp trawling operations is a problem in some areas but is being widely addressed through the use of Turtle Excluder Devices (TEDs) which operate on similar principles to other bycatch reduction devices.

BEAM TRAWLING

The ecosystem effects of beam trawling are to a large extent comparable to those of demersal otter trawls. Compared with otter trawls, however, beam trawls will generally have a poorer energy efficiency and a stronger impact on bottom habitats.

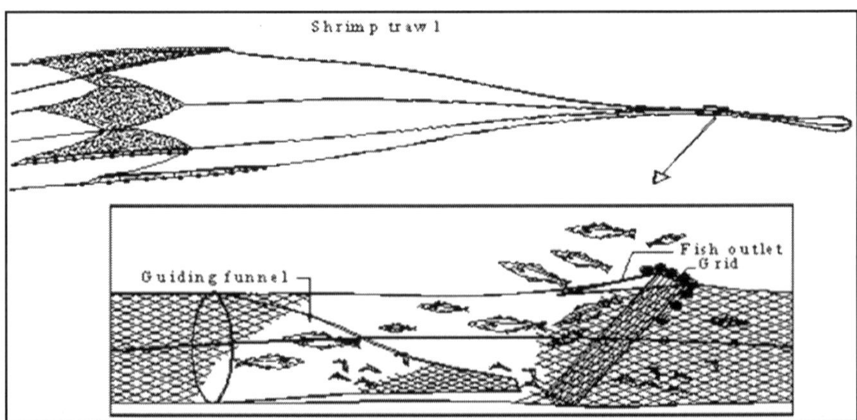

Fig. Shrimp trawl with sorting grid (expanded view). Shrimp and fish that pass backwards in the trawl are guided by a funnel to the bottom of the backwards slanting metal grid. Shrimp and fish of comparable size will pass through the slots of the grid and go to the cod end, while larger fish and other organisms (e.g. jellyfish) will slide upwards over the grid and are released through the outlet.

SEINE NETS

Seine netting is also fairly similar to demersal trawling with respect to most ecosystem effects. However, the seine net is considered to cause less habitat destruction, is more energy efficient and does produce a better catch quality overall.

PURSE SEINE

Purse seining is a non-selective gear regarding fish size, as the mesh size is chosen to be so small that there should be no risk of mass meshing of fish, even by the smallest size groups of the target species. However, in cases where the fish size in the catch is too small, as estimated from samples taken from the seine, there is usually an opportunity to release the fish. The species selectivity is fairly high and both from the fishers experience and by use of modern sonar equipment it is not too difficult to identify the species before the seine is set. There is a certain risk of by-mortality in purse seining. Pelagic fishes are in general sensitive to contact with fishing gears which easily leads to loss of scales and resulting mortality. This can be related to the above-mentioned release of unwanted species or sizes of fish, but the main cause of by-mortality in purse seining is the escapement of fish after net rupture due to large catches and/or bad weather.

There is extremely low risk of ghost fishing with lost purse seines. The energy efficiency is high because of the relatively large catches that give a high catch-per-unit-effort in this fishery. Catch quality is normally also high, particularly in modern purse seining, where the catch is pumped directly into refrigerated tanks on the fishing vessel. Purse seining has generated some adverse publicity as a result of bycatches of dolphin in some tuna fisheries, but effective methods to avoid such capture have been developed.

BEACH SEINES

Beach seines have poor selectivity properties, catching a variety of species and sizes of fish and other organisms. There may be some by-mortality associated with beach seining, while the energy efficiency and the catch quality of this gear is generally high.

MANAGEMENT CONSIDERATIONS: SELECTIVITY AND OTHER ECOSYSTEM EFFECTS OF FISHING

An example of how the properties of different fishing gears could be evaluated in terms of their selectivity and ecosystem effects. Here the various ecosystem effects are given a rank from 1 (non-favourable) to 10 (favourable), giving an overall index of average ecosystem effect. This table must of course only be regarded as a guideline and an example of how to approach an evaluation of different fishing gears and fisheries from a management point of view.

Thereafter the specific fishery in an area should be analysed in more detail and the proposed ecosystem factors should also be weighted according to their importance in a local or regional case. Although a fishing method might be characterised as being more or less responsible in general, attention should be paid to where, when and how it is being used. The evaluation and elaboration of technical regulations should be done in co-operation with the fishers to establish better understanding by them of the aim of the regulations and to hear and consider their advice on the regulations and their implementation.

Table. Generalized estimate of ecosystem effects of fishing for different fishing methods - ranked on a scale from 1 (non-favourable) to 10 (highly favourable) with respect to different ecosystem related factors.

Ecosystem effects and Gear type	Size selection	Species selection	By-mortality	Ghost fishing	Habitat effects	Energy efficiency	Catch quality	Ecosystem effect index
Gillnets	8	4	5	1	7	8	5	5,4
Trammel nets	2	3	5	3	7	8	5	4,7
Handlining	4	4	6	10	9	9	9	7,3
Longlining	6	5	6	9	8	8	8	7,1
Pots	7	7	9	3	8	8	9	7,3
Traps	5	5	8	8	9	9	9	7,6
Spear, harpoon	8	9	5	10	10	8	9	8,4
Pelagic trawl	4	7	3	9	9	4	8	6,3
Demersal trawl	4	4	6	9	2	2	6	4,7
Beam trawl	4	4	6	9	2	1	6	4,6
Shrimp trawl	1	1	7	9	4	2	6	4,3
Seine net	5	5	6	9	4	5	8	6,0
Purse seine	-	7	5	9	9	8	8	7,7
Beach seine	2	2	5	10	6	9	9	6,1

Other specific factors, for example their socio-economic implications, could be added to a specific evaluation, as a guideline to future management strategies with respect to choice and priorities between different fishing methods. These factors should be included with the information used to assist in the design of management strategies. This form of evaluation can also be used to identify present or future ecosystem-related weaknesses with existing fishing methods and practises as a baseline for research and development aiming at improvements of selective properties and undesired ecosystem effects.

Bibliography

A.C. Long: *A-Z of Industrial Marine Fisheries in Asia*, Cyber Tech Publications, New Delhi, 2011.

A.D. Dholakia: *Fisheries and Aquatic Resources of India*, Daya Publication, Delhi, 2004.

A.K. Sarkar, R.S. Singh, M.S. Yadav and C.S. Singh: *Integrated Farming Systems for Sustainable Production*, Agrotech Publishing Academy, Delhi, 2011.

A.P. Jaiswal, N.C. Gupta and R.B. Sah: *Crop Management and Integrated Farming*, Enkay Publishing House, Delhi, 2011.

A.St.J. Macdonald: *Circumventing the Mahseer and Other Sporting Fish in India and Burma*, Natraj Publications, Delhi, 2001.

Ajit Kumar Roy: *Evaluation and Impact Assessment of Technologies and Developmental Activities in Agriculture, Fisheries, and Allied Fields*, New India Publishing Agency, Delhi, 2011.

Albert C L G Gunther and J Arthur Thomson: *Biology of World Fishes*, Arise Publishers, Delhi, 2012.

Amita Saxena: *Fisheries Economics*, Daya Publishing House, Delhi, 2011.

B.N. Pandey, Sunil P. Trivedi, Kamal Jaiswal and Navneet Kaur Sethi: *Fish and Fisheries*, Sarup Book Publishers, Delhi, 2009.

B.R. Selvamani and R.K. Mahadevan: *Fish and Fishery Culture : Assessment and Evaluation*, Campus Books International, Delhi, 2008.

B.R. Selvamani and R.K. Mahadevan: *Fish Farming Systems*, Campus Books International, Delhi, 2008.

B.R. Selvamani and R.K. Mahadevan: *Freshwater Fish Farming*, Campus Books International, Delhi, 2008.

Bansi Lal Kaul: *Advances in Fish and Wildlife Ecology and Biology, Vol. II*, Daya Publications, Delhi, 1999.

C.B.L. Srivastava: *A Text Book of Fishery Science and Indian Fisheries*, Kitab Mahal, Delhi, 2013.

D.N. Mishra: *Aquatic Resources and Fisheries*, Surendra Publication, Delhi, 2010.

D.R. Khanna and P.R. Yadav: *Biology of Fishes*, Discovery Publications, Delhi, 2004.

Harish Lowe: *Beginner's Guide to Aquarium Fish and Fish Care*, Abhishek Press, Delhi, 2005.

Jagdish Ojha: *Biology of Hillstream Fishes*, Narendra Publications, Delhi, 2002.

K.D. Bhardwaj: *A-Z of Fisheries and Aquaculture Technology*, Cyber Tech Publication, New Delhi, 2011.

K.D. Bhardwaj: *Business Management in Fisheries and Aquaculture*, Cyber Tech Publication, New Delhi, 2012.

N.M. Chakrabarty, P.P. Chakraborty and S.C. Mondal: *Biology Breeding and Farming of Important Food Fishes*, Narendra Publications, Delhi, 2010.

Nasreen Akhter: *Economics in Fisheries Research*, Oxford Book, Delhi, 2010.

Neha Charan: *Classification of Fisheries*, Random Publications, Delhi, 2013.

Q. J. Shammi and S. Bhatnagar: *Applied Fisheries*, Agrobios Publication, Jaipur, 2002.

Rajendra Kumar: *A Textbook of Fish and Fisheries*, Arise Publication, Delhi, 2008.

Rajiv Tyagi and Arvind N. Shukla: *Development of Fishes*, Anmol Publications, Delhi, 2002.

Rashiduzzaman Ahmed: *Fish and Wildlife of Arua and Gopinathpur Beel Areas*, IUCN-The World Conservation Union, 2005.

Richard D. Fisher: *China's Military Modernization: Building for Regional and Global Reach*, Pentagon Press, Delhi, 2009.

V P Saini; L L Sharma and N C Ujjania: *Dictionary of Aquatic Resources and Fisheries*, Agrotech Publications, Rajasthan, 2007.

V. Suryanarayan: *Conflict Over Fisheries in the Palk Bay Region*, Lancer Publication, Delhi, 2005.

V.B. Sakhare and B. Vasanthkumar: *Emerging Trends in Fisheries and Aquaculture*, Daya Publication, Delhi, 2013.

V.B. Singh: *Fish Farming*, ALP Books, Delhi, 2010.

Index